T0265935

Cambridge Primary

Mathematics

Second Edition

Learner's Book 4

Steph King
Josh Lury
Series editors:
Mike Askew
Paul Broadbent

Acknowledgements

The Publishers would like to thank the following for permission to reproduce copyright material.

Photo credits
p. 22 *cr* ©Hachette UK; **p. 35** *b* © Andrii Vergeles/Adobe Stock Photo; **p. 40** *r* © Lucas Vallecillos/VWPics/Alamy Stock Photo; **p. 62** *tr* © Red Confidential/Shutterstock.com; **p. 76** *cc* ©Hachette UK; **p. 76** *cc* © Atoss/Adobe Stock Photo; **p. 106** *bl* © Fairyn/Adobe Stock Photo; **p. 106** *br* © Valerie Potapova/Adobe Stock Photo; **p. 139** *tl* ©Hachette UK; **p. 168** *t* © Rawpixel.com/Adobe Stock Photo.

t = top, *b* = bottom, *l* = left, *r* = right, *c* = centre

Orders: please contact Hachette UK Distribution, Hely Hutchinson Centre, Milton Road, Didcot, Oxfordshire, OX11 7HH. Telephone: +44 (0)1235 827827. Email education@hachette.co.uk Lines are open from 9 a.m. to 5 p.m., Monday to Friday. You can also order through our website: www.hoddereducation.com

ISBN: 978 1398 301 02 3
© Steph King, Josh Lury 2021
First published in 2017
This edition published in 2021 by
Hodder Education,
An Hachette UK Company
Carmelite House
50 Victoria Embankment
London EC4Y 0DZ
www.hoddereducation.com

Impression number   10 9 8 7 6 5 4 3 2 1
Year   2025 2024 2023 2022 2021

Cover illustration by Lisa Hunt, The Bright Agency

Illustrations by James Hearne, Natalie and Tamsin Hinrichsen, Ammie Miske, Vian Oelofsen

Typeset in FS Albert 17/19 by IO Publishing CC

Printed in Italy

A catalogue record for this title is available from the British Library.

Contents

How to use this book

This book will help you to learn about mathematics.

Explore the picture or problem.
What do you see?
What can you find?

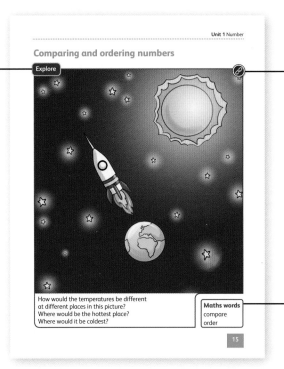

This icon shows you that the activity links with other subjects in your school curriculum.

Understand new **Maths words**.
The *Mathematical dictionary* at the back of this book can help you.

Learn new mathematics skills with your teacher.
Look at the pictures to help you.

Do the Practise activities to learn more.
Work like a mathematician.

The shaded questions show you what you need to do.

Remember to write any answers in your notebook, not in this textbook.

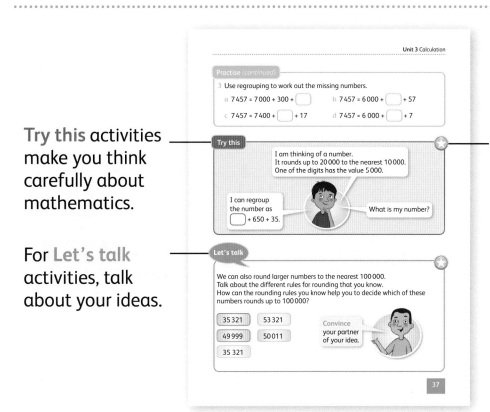

Try this activities make you think carefully about mathematics.

For **Let's talk** activities, talk about your ideas.

This star shows you the activities that require you to think and work mathematically. This means:
- **specialising** and **generalising**
- **conjecturing** and **convincing**
- **characterising** and **classifying**
- **critiquing** and **improving**.

Do each **Quiz** to find out how much you have learnt.

This icon shows you that audio material is available. Listen and you will learn.

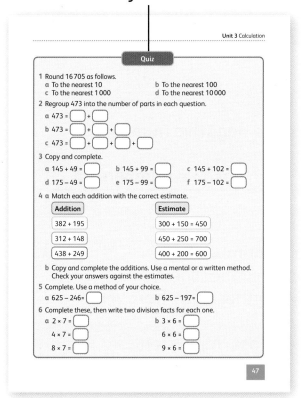

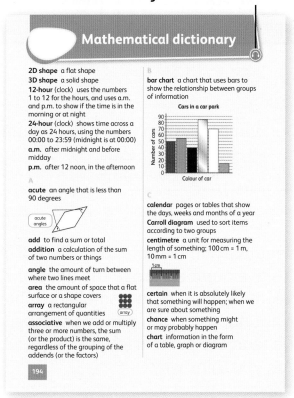

Negative numbers

Guss and Elok are putting numbers on this number line.

What do you notice about the number line?

Find the missing numbers.

Maths words
positive number
negative number
zero
number line
explain

Learn

Our number system uses **positive numbers** and **negative numbers**.

negative numbers   positive numbers

0

Negative numbers are to the left of **zero** on a **number line**.
Positive numbers are to the right of zero on a number line.
Negative numbers get smaller as they move further away from zero.
What happens with positive numbers?
We read the number −10 as **negative ten** and not 'minus ten'.
Can you think why?

Practise

1 Read and say these numbers. Is each number to the left or right of zero?

−5	−25	−12	25	−34

2 a Write the number that replaces each letter on this number line. Choose numbers from the box.

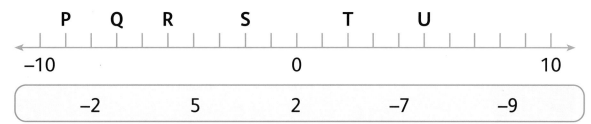

−2	5	2	−7	−9

b One letter is left over. What number does it represent?

3 Copy and complete. Write **closer to** or **further away from**.

a −20 is _____ zero than −12

b −2 is _____ zero than −8

c −5 is _____ zero than 4

d 6 is _____ zero than −7

Let's talk

Guss has put negative and positive numbers on this number line.

−1	−4	−7	**0**	9 8	5	2

Use the skills of **critiquing** and **improving** to **explain** the mistakes Guss has made.

Counting on and back

What different counting patterns can you find?

1031	250	450	393
931	350	650	
405	831	550	403
410	731	413	
419	415	631	423
420	427	531	
421	425	429	433
423	430	431	

Maths words
tens
hundreds

Convince a partner that your counting patterns work.

Learn

We can count on and back in different steps.
Count on in **tens** from 546. What do you notice?
Now count on in **hundreds** from 546.
What do you notice this time?

Let's count back from 5 in steps of three.
What numbers do you say?

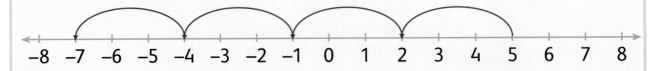

Practise

1 Write the first five numbers you say in each count.

   a Count back in fours from 8.   b Count on in fives from −16.

   c Count back in tens from 512.   d Count on in tens from 476.

   e Count back in hundreds from 999.   f Count on in hundreds from 684.

2 Look at these numbers.

  (497)   (552)   (597)   (648)   (352)   (507)   (512)

   a Which numbers will you say when you:

      • count on in steps of 5 from 432?

      • count back in steps of 10 from 617?

      • count back in steps of 100 from 1 052?

   b Which number from above has not been used?
     Complete these sentences to show two counts this number will be in.

     Count on in steps of 10 from _____ .

     Count back in steps of 100 from _____ .

Try this

Banko is counting back from 10.
He chooses one of these numbers as the step to count back in:

  [6]   [4]   [3]

He says the number **negative eight** in his count.
Which number did Banko choose? Is there more than one answer?
What else do you notice about the different numbers in each count?

Number and place value

Explore

How many small cubes make a flat square?

How many small cubes make a larger cube?

How many flat squares make a larger cube?

Choose an equal number of small cubes, sticks, flat squares and larger cubes.

What different totals can you make?

For each of your totals, take away one piece.

How does it change the total each time?

Maths words
digit
thousands
decompose
place value

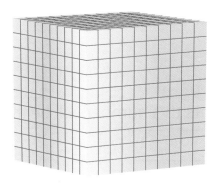

Learn

Each **digit** in a number represents a different value.
The digit 2 appears twice in the number 2325.
The first digit 2 represents two **thousands**.
The second digit 2 represents two tens.

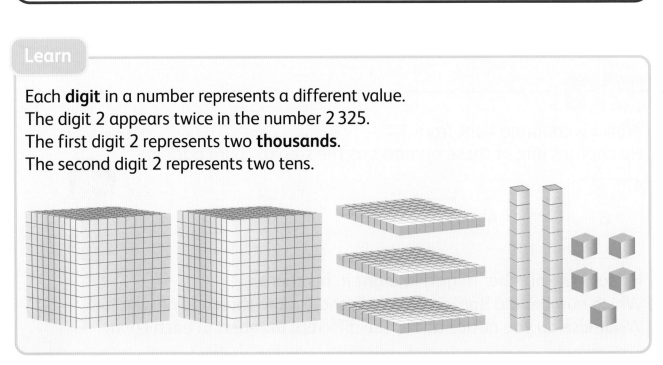

Learn *(continued)*

We write 2 325 in words as **two thousand, three hundred and twenty-five**.
The number sentence 2 325 = 2 000 + 300 + 20 + 5 shows how we **decompose**
2 325 into **place value** parts.
How do you know there are 23 hundreds in 2 325?
How do you know there are 232 tens in 2 325?
What happens when you count on from 2 325 in one step of 1 000?
What will the number be now?

Practise

1 Write these numbers as numerals.   One thousand and one = ⎣1 001⎦

   a One thousand and ten = ⎣ ⎦

   b One thousand, one hundred and ten = ⎣ ⎦

   c Two hundred and twenty-two = ⎣ ⎦

   d One thousand, two hundred and thirty-four = ⎣ ⎦

   e Four thousand, three hundred and twenty-one = ⎣ ⎦

2 Write number sentences to decompose each number into place value parts.

   ⎣4 305 = 4 000 + 300 + 0 + 5⎦   a 4 035 = ⎣ ⎦

   b 3 450 = ⎣ ⎦   c 3 432 = ⎣ ⎦

   d 4 433 = ⎣ ⎦   e 5 324 = ⎣ ⎦

Practise *(continued)*

3 Copy and complete.

a 1 000 more than 3 452 is ☐ .

b 1 000 less than 1 523 is ☐ .

c When we count on in thousands from 4 608, the next two numbers

we say are ☐ and ☐ .

4 Complete these part-whole diagrams.

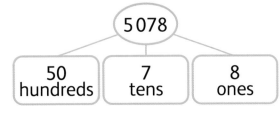

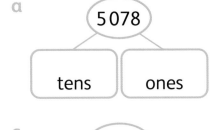
a

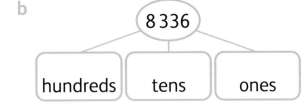
b

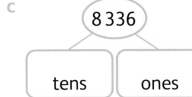
c

Let's talk

Use your skills of specialising and generalising.
What different 4-digit numbers can you make that have:

- 36 hundreds?
- 240 tens?

Can you explain why it is possible
to make more 4-digit numbers with
36 hundreds than with 240 tens?
Convince your partner.

Learn

The number **35 374** has five digits. Look at this place value chart.

10 000s	1 000s	100s	10s	1s
3	5	3	7	4

The digit **3** appears twice in 35 374.
What is the value of each digit 3?
We read 35 374 as **thirty-five thousand, three hundred and seventy-four**.
How many thousands does 35 374 have?
How many hundreds?

Practise

1 Read and write these numbers in numerals.

a Seventeen thousand, six hundred and twenty-three =

b Twenty-seven thousand, three hundred and twenty-six =

c Seven thousand, two hundred and sixty-three =

d Seventy-two thousand and sixty-three =

2 Write the value of each underlined digit.

a 2<u>8</u> 057

b <u>8</u>2 057

c 5<u>4</u> 325

d <u>54</u> 532

e <u>70</u> 2<u>7</u>1

f 72 <u>2</u>17

You could use a place value chart like the one in Learn to help you make decisions.

Practise *(continued)*

3 Answer these.

> How many more thousands in 45 500 than in 39 500? | 6 thousands |

a How many more thousands in 23 450 than in 19 450? [   ]

b How many more hundreds in 32 400 than in 31 300? [   ]

c How many more hundreds in 41 835 than in 39 735? [   ]

Let's talk

Use your skills of critiquing and improving.

We read 42 385 as **forty-two thousand, three hundred and eighty-five.**

No, we read 42 385 as **four ten thousands, two thousand, three hundred and eighty-five.**

Who is correct, Banko or Pia?
Explain the mistake the other child has made.
How can you help that child to improve in reading numbers?

Comparing and ordering numbers

Explore

How would the temperatures be different
at different places in this picture?
Where would be the hottest place?
Where would it be coldest?

Maths words
compare
order

Learn

Look at these thermometers. What is the same and what is different about them?

We can **compare** the two temperatures as 15 °C **>** –5 °C and –5 °C **<** 15 °C.
15 °C is warmer than –5 °C. Temperatures above 0 °C are warmer than temperatures below 0 °C.

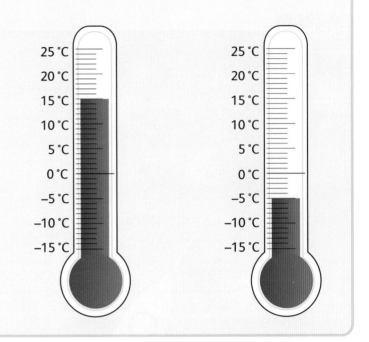

Practise

1 Write the temperature shown each time.

a 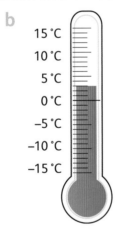   b 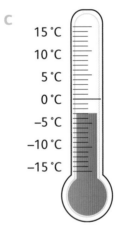   c   d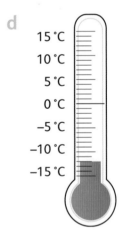

2 Which temperature in each pair is colder?

   a 3 °C or –3 °C   b –4 °C or –3 °C   c 0 °C or 12 °C

   d 16 °C or –2 °C   e –2 °C or –12 °C

3 Use the **<** or **>** symbol to show what you found out about comparing the temperatures in question 2.

Try this

Use your skills of critiquing and generalising.

Twenty degrees Celsius is warmer than fourteen degrees Celsius because it is further away from zero degrees Celsius.

Minus twenty degrees Celsius is warmer than minus fourteen degrees Celsius because it is also further away from zero degrees Celsius.

Who do you agree with, Jin or Guss? Why?
Can you make up a rule to help them compare any pair of temperatures?
What do you need to tell them?

Learn

We can use the symbols =, > and < to show how larger numbers compare.
It is important to know the place value of each digit first.

	10 000s	1 000s	100s	10s	1s
12 435	○	○○	○○ ○○	○○ ○	○○ ○ ○○
12 345	○	○○	○○ ○	○○ ○○	○○ ○ ○○

12 435 and 12 345 both have 12 thousands, so we must look at
the hundreds next.
What do you notice?
We can write 12 435 > 12 345 to show that 12 435 is the larger number.
How can we compare the numbers using the < symbol?

Practise

1 Use the symbol **<** or **>** to make each sentence true.

a 7 325 ☐ 7 235

b 5 033 ☐ 5 303

c 15 033 ☐ 13 335

d 32 782 ☐ 23 872

e 14 423 ☐ 14 432

2 True or false? Correct any that are false.

a 8 324 > 8 342

b 3 240 < 999

c 3 240 = 324 tens

d 12 049 > 12 039

e 17 600 = 167 hundreds

f 19 945 < 19 899

3 Write these numbers in order from smallest to largest each time.

a 5 436 5 463 654 5 346

b 6 402 6 042 6 420 6 204

c 21 304 12 304 13 240 21 034 13 402

Let's talk

Answer these. **Convince** your partner of your decisions.

Which number is closest to 10 000?	Which number is closest to 8 000?	Which number is closest to 7 400?	Which number is closest to 6 230?
11 300	7 600	7 372	6 215
9 400	8 532	7 430	6 242
10 700	8 329	7 299	6 195

Order the first column of numbers from smallest to largest.
Include the number 10 000.

Rounding numbers

Explore

Elok and Pia are putting numbers on the number lines.

Maths word
round

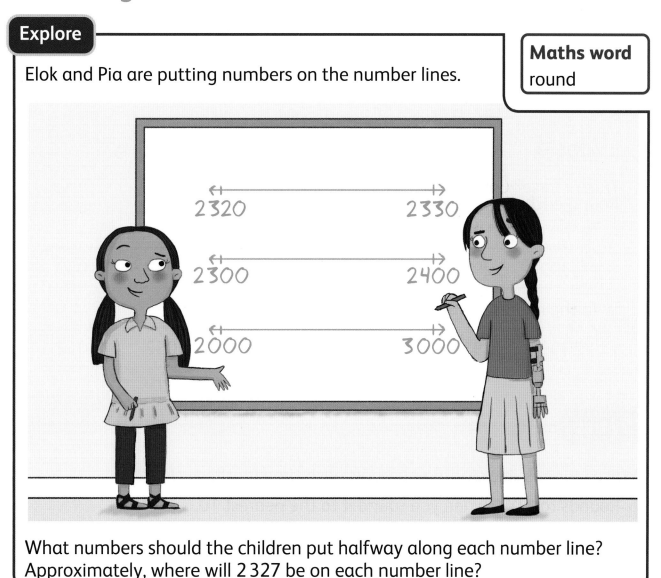

What numbers should the children put halfway along each number line?
Approximately, where will 2 327 be on each number line?

Learn

2 327 **rounds** up to 2 330 to the nearest 10 because it is closer to
2 330 than 2 320.
Numbers with 5 or more in the ones position round up to the nearest 10.
Does 2 327 round up or round down to the nearest 100? Why?
What digit should we check when rounding to the nearest 100?
We check the hundreds digit when we round to the nearest 1 000.

Practise

1 Round to the nearest 10.

234 → 230

a 295 → ▢

b 3 434 → ▢

c 4 295 → ▢

d 5 702 → ▢

e 5 726 → ▢

2 Round each number as follows.
a To the nearest 100

| 2 309 | 3 291 | 4 395 | 2 849 | 2 463 | 3 052 | 13 052 |

b To the nearest 1 000

| 2 309 | 3 291 | 4 395 | 2 849 | 2 463 | 3 052 | 13 052 |

Try this

Write four possible starting numbers each time for these clues.

a Rounds to 1 000 when you round it to the nearest 10.

b Rounds to 2 200 when you round it to the nearest 100,
and to 2 240 when you round it to the nearest 10.

c Rounds to 2 800 to the nearest 100, and to 3 000 to the nearest 1 000.

d Rounds to 3 200 when you round it to the nearest 10.

I wonder what the largest possible
starting number will be each time?

Quiz

1 What is the value of each letter?

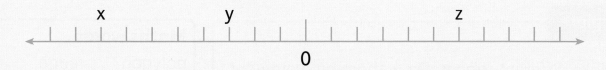

2 Count on in tens from 689. Write the first three numbers you say.

3 Count back in hundreds from 235. Write the first three numbers you say.

4 What is the value of the digit(s) 7 in each number?
   a 2374
   b 7234
   c 17234
   d 71734

5 True or false? Correct any that are false.
   a 5324 = 5000 + 300 + 20 + 4
   b 3402 = 3000 + 40 + 0 + 2
   c 4065 = 4000 + 600 + 5

6 Write =, > or < to make each sentence true.
   a 4576 ⬚ 5476
   b 15476 ⬚ 14576

7 Write the warmer temperature each time.
   a −11 °C or 5 °C
   b 11 °C or −15 °C

8 Round the number 15654 as follows.
   a To the nearest 10
   b To the nearest 1000

Polygons

Explore

Make different shapes. Draw them on a nine-dot grid or use elastic bands and a nine-pin geoboard.

Maths words

polygon	edge
angle	right angle
parallel	vertices

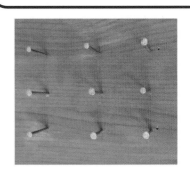

How many different triangles can you make?
What types of triangles are they?
Make some quadrilaterals (four-sided shapes).
Try to name them all.
Make a shape with more than four edges.

Learn

In a regular **polygon**, all the **edges** and all the **angles** are equal.
Name each shape. Discuss with a partner where to place it in the sorting diagram.

a   b   c

d 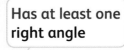   e   f

Has at least one right angle   Polygon

g   h   i

Irregular polygon   Has **parallel** edges

j   Try to name other headings to sort the shapes in different ways.

Hexagon   Regular polygon

Quadrilateral   Triangle

Practise

1 Look at these polygons.

A   B   C

D   E   F

a Which shapes are hexagons?

b Which shape is a hexagon with three right angles?

c Which shape is a pentagon with one right angle?

d Which hexagon has the most right angles?

e Are there any shapes with an even number of right angles?

2 Draw your own shapes on a nine-dot grid. Use your skill of specialising and generalising to sort them according to the criteria in the diagram.

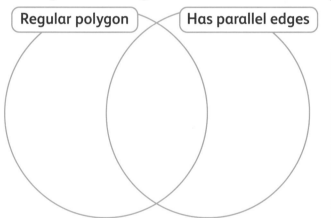

Regular polygon   Has parallel edges

Let's talk

Is it possible to make a shape that is not a polygon on a nine-pin geoboard? Discuss and explain.

Try this

How many different shapes are hidden in this diagram?
Can you see different hexagons or pentagons? Draw them.
What shape has the most **vertices**?

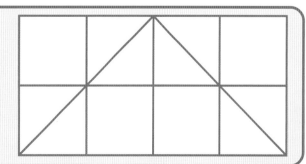

Compound shapes and tessellation

Explore

The tiles in this pattern **tessellate**.
This means that they join without leaving gaps.
Could this pattern continue forever?

Which of these shapes do you think will tessellate?

Maths words
tessellate
compound shape

Learn

We can make **compound shapes** by joining two or more simple shapes. We made these compound shapes by joining three polygons (two squares and a rectangle) in different ways. We joined the edges of the polygons to make one large rectangle and one L-shape.

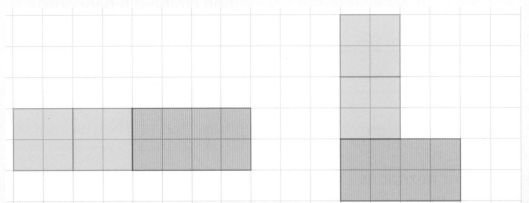

Try to make other compound shapes using three polygons like this.

Practise

1 Cut a square into the two shapes shown.

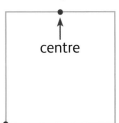 → 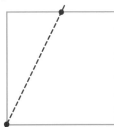 →

centre

Join the shapes to make different compound shapes.
Sketch the compound shapes you make.

2 Use two squares. Cut each square as shown.

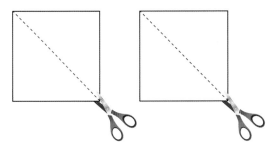 

You should have four identical triangles.

a Join two of the triangles to make:
   • a triangle
   • a four-sided shape that is not a square.

b Join all four triangles to make:
   • a rectangle
   • a square
   • a triangle
   • a different quadrilateral.

c What shapes can you make by joining three triangles?

3 Many copies of this shape are used to make a pattern.
   a Does the shape tessellate?
   b What do you notice about the pattern?
   c Explore tessellating your own 2D shapes.
   Try to find some shapes that tessellate and some that do not.

Compound and irregular shapes

Explore

Jin drew an irregular shape. It has sides of different lengths.
He worked out its area.

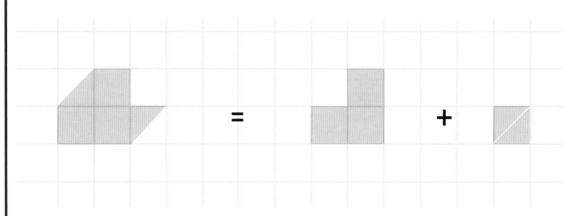

My shape has an area
of four square units.

Discuss Jin's method for working out the area.

Jin measures the area in square units.

If each square on the grid is 1 cm by 1 cm, the units are called
square centimetres.

We can write this as cm^2.

We can also use larger units. Discuss m^2 and km^2.

What would you use these units to measure?

Maths word
compound shape

Learn

Each square on this grid is 1 cm by 1 cm.

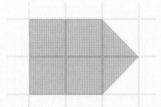

To find the area of the shape, we can count the orange squares and use Jin's method to join half squares: $4 + \frac{1}{2} + \frac{1}{2} = 5 \, cm^2$.

Another method is to break the shape into two or more regular shapes.

The shape is made up of a square and a triangle.

Together, these shapes make a **compound shape**.

It is more difficult to find the area of irregular shapes with curved edges.

We can only estimate the area.

We count the whole squares first.

Then we count the squares that are more than half covered as a whole square.

Ignore the squares that are less than half covered.

What is your estimate for the area of this shape?

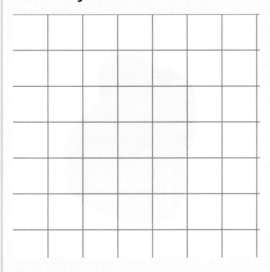

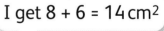
I get 8 + 6 = 14 cm²

Is Banko correct?

1 Work out the area of each shape in square units.
Try drawing your own shapes like these on a square grid.

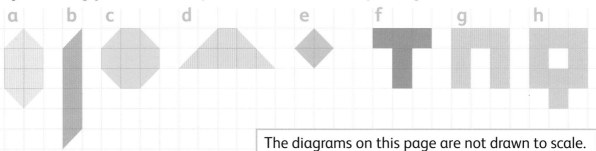

> The diagrams on this page are not drawn to scale.

2 Work out the area of each compound shape.

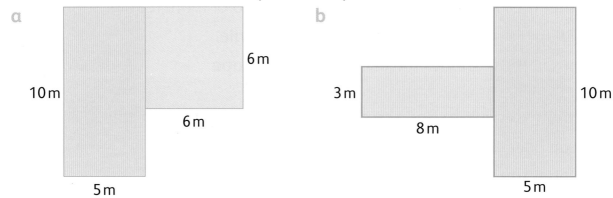

⭐ 3 These squares are drawn on centimetre square grids.

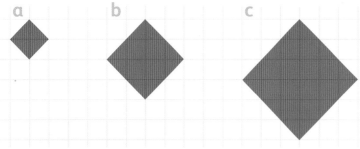

Use your conjecturing skills to estimate the area of each shape, then check. Write the area in cm².

4 Estimate the area of each pond.

Justify your reasoning.

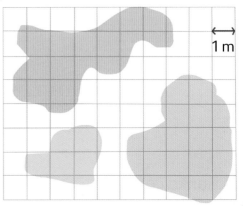

Squares and rectangles

Explore

Estimate the width and length of each shape.
Will the measurements be in **millimetres**,
centimetres or metres? Measure to check.

Maths words

millimetre
centimetre
area
perimeter
straight

Learn

Jin thinks the **area** of this rectangle is 20 cm². Let's investigate.

9 cm

2 cm

To check, we can draw the rectangle on a square grid.
Each square on the grid is 1 cm by 1 cm.
We can use this fact to work out the area of the rectangle.

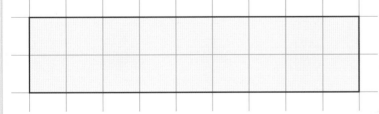

Counting all the squares in the
rectangle will tell us the area.
Count them.

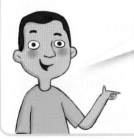

I can count all the
squares one by one.

I can see a way to
calculate the area
without counting.

Was Jin's estimate over or under the actual
area of the rectangle? By how much?

Practise

1 Predict which shape has the largest area in cm².
Then use your ruler to check.

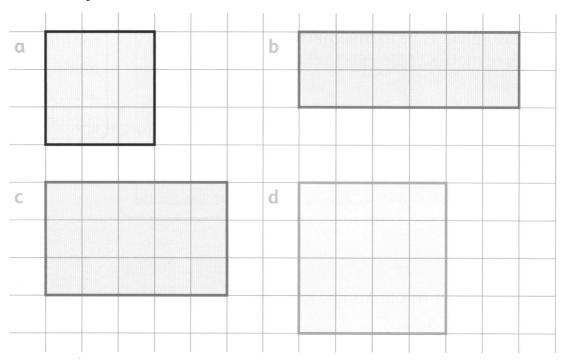

a
b
c
d

2 Use a ruler to draw each rectangle. Then work out how many centimetre
squares each rectangle covers. Write your answers in cm².

a Width = 5 cm, length = 5 cm
b Width = 4 cm, length = 3 cm
c Width = 4 cm, length = 6 cm
d Width = 10 cm, length = 4 cm
e Width = 8 cm, length = 3 cm
f Width = 2 cm, length = 12 cm

 3 Draw these shapes on a 1 cm by 1 cm grid.

a A square with an area of 49 cm²
b A rectangle with an area of 30 cm²
c A rectangle with an area of 12 cm²

There may be more than one solution for some of these challenges.
Convince a partner that your solution is correct.

Learn

Look at the two rulers.
One shows centimetres.
The other shows millimetres.
Millimetres are very small
measurements.
There are 10 mm in 1 cm.

Perimeter is the distance all the way
around the outline of a shape.
Perimeter is given in units of distance,
for example, centimetres or metres,
the same as when you measure a
straight line.
The ant walks around the perimeter
of this shape in straight lines.
5 cm + 3 cm + 5 cm + 3 cm = 16 cm
In total, the ant walks 16 cm.

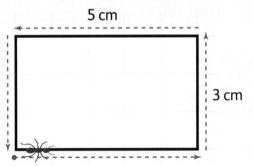

Practise

1 Measure the perimeter of each shape. Use a ruler.

a

Give your answer
in centimetres.

b

Give your answer
in centimetres.

Practise *(continued)*

c

d

Give your answer in millimetres.

Give your answer in millimetres.

2 Calculate the perimeter of each shape.

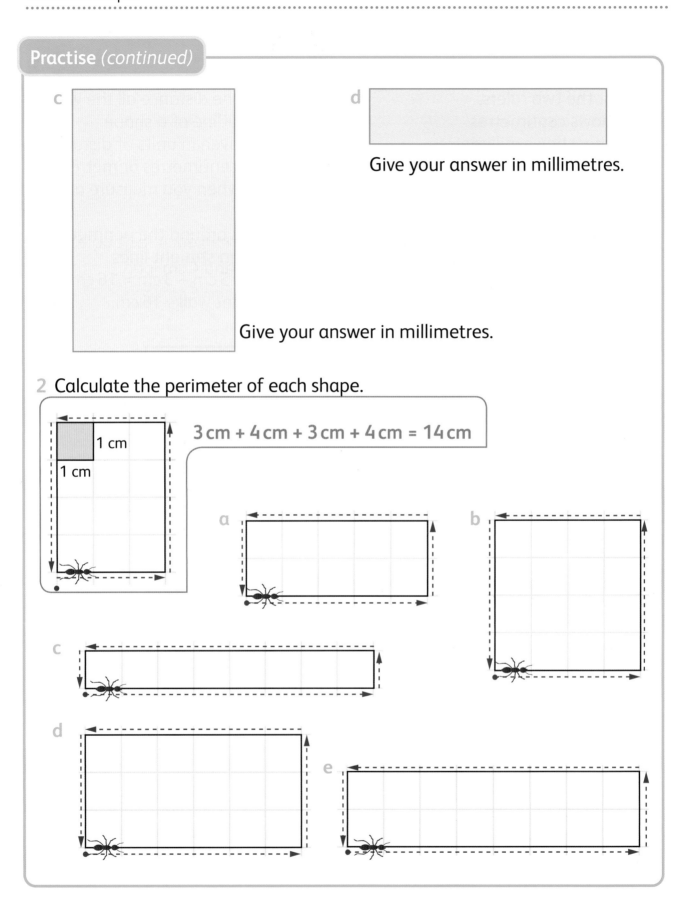

$3\,cm + 4\,cm + 3\,cm + 4\,cm = 14\,cm$

1 cm

1 cm

a

b

c

d

e

Practise *(continued)*

3 Use your specialising skills. Estimate the perimeter of each of these.
  a Your classroom   b Your desk
  c The school playground   d This textbook
  Think carefully about the units of measurement you choose.
  Now ask your teacher if you can take the exact measurements.

4 Predict the perimeter of each shape. Check by drawing each rectangle.
  a 5 cm wide and 3 cm long   b 6 cm wide and 4 cm long
  c 7 cm wide and 2 cm long   d 6 cm wide and 6 cm long

Try this

Try drawing shapes with these perimeters:
a a rectangle with a perimeter of 30 cm
b a triangle with a perimeter of 30 cm
c a square with a perimeter of 30 cm.

Use your skills of improving.
Which shape is easiest to draw?
Which shape is the most difficult? Why?

Let's talk

Draw shapes with these perimeters.

a 10 cm   b 20 cm   c 100 mm

Check your drawings with your partner's drawings.
Are your shapes the same? Use your improving skills.

Quiz

1 Write two headings that could have been used to sort these shapes.

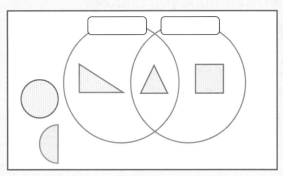

2 Describe, name or draw different compound shapes that can be made by joining two or more copies of a triangle like this. Use any size you like.

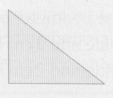

3 What is the area of a square with sides of 10 m?

4 Which ant has further to walk?

a

90 mm

10 mm

b

4 cm

5 cm

5 Work out the area of these shapes. Answer in square units.

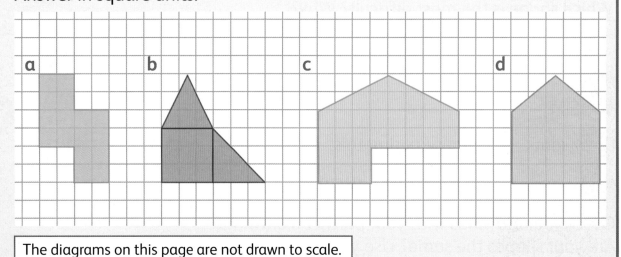

a   b   c   d

The diagrams on this page are not drawn to scale.

Skills for calculating

Explore

Read the heights of these mountains.
Which heights **round** to 9 000 m to the **nearest** 1 000?
Which height rounds to 8 500 m to the nearest 100?
What other rounding can you do?
Approximately how many metres higher is Makalu
than Broad Peak?

Maths words
round
nearest
decompose
regroup

Mountain	Height (m)
Mount Everest	8 848
K2	8 611
Makalu	8 485
Broad Peak	8 051
Nanda Devi	7 816

Learn

Knowing the place value of individual digits can help you to round and **decompose** numbers. The height of the mountain called Makalu is 8 485 m. 8 485 is represented in this place chart.

1 000s	100s	10s	1s
⊙ ⊙ ⊙ ⊙ ⊙ ⊙ ⊙ ⊙	⊙ ⊙ ⊙ ⊙	⊙ ⊙ ⊙ ⊙ ⊙ ⊙ ⊙ ⊙	⊙ ○ ⊙ ⊙ ⊙

We can **decompose** 8 485 to show the value of each digit.

8 485 = 8 000 + 400 + 80 + 5

We can also **regroup** 8 485 in different ways:

8 485 = 8 000 + 485
8 485 = 8 000 + 400 + 85
8 485 = 7 000 + 1 400 + 85

Regrouping numbers is an important skill when calculating. Why do you think this is?

Practise

1 Decompose these numbers to show the value of each digit.

   a 7 457   b 8 504   c 11 085   d 15 142

2 Copy and complete this table.

	Rounded to the nearest 10	Rounded to the nearest 100	Rounded to the nearest 1000	Rounded to the nearest 10 000
7 457				
8 504				
11 085				
15 142				

Practise *(continued)*

3 Use regrouping to work out the missing numbers.

a 7 457 = 7 000 + 300 + ▢

b 7 457 = 6 000 + ▢ + 57

c 7 457 = 7 400 + ▢ + 17

d 7 457 = 6 000 + ▢ + 7

Try this

I am thinking of a number.
It rounds up to 20 000 to the nearest 10 000.
One of the digits has the value 5 000.

I can regroup
the number as
▢ + 650 + 35.

What is my number?

Let's talk

We can also round larger numbers to the nearest 100 000.
Talk about the different rules for rounding that you know.
How can the rounding rules you know help you to decide which of these numbers rounds up to 100 000?

35 321	53 321
49 999	50 011

35 321

Convince
your partner
of your idea.

Using rounding to help with adding and subtracting

The children are playing a game. Play the game with a partner.
Choose one number from the box and another number from the hoop to **add** or **subtract**.
Make up some calculations for each other.
Estimate the answers.
How can you use rounding to help you estimate?

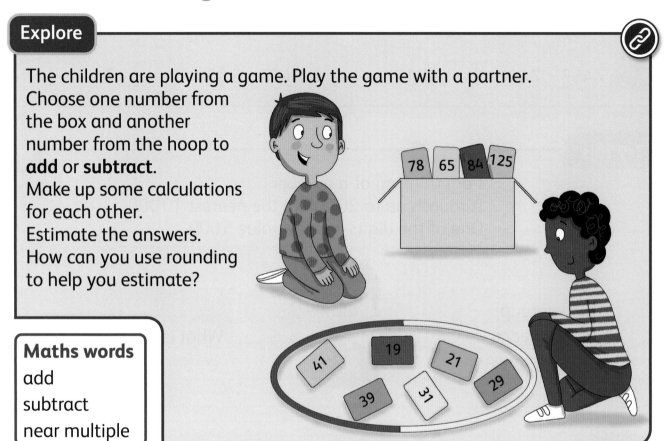

Maths words
add
subtract
near multiple

Learn

These number lines show the two calculations **78 – 29** and **78 – 31**.
The numbers 29 and 31 are **near multiples** of 10. They both round to 30.

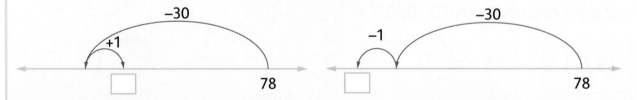

We can use what we know about place value to help us subtract tens.
We can write 78 – 29 as 78 – 30 + 1.
Why do we need to add 1 when we subtract 29 in this way?
We can write 78 – 31 as 78 – 30 – 1.

Practise

1 Solve these pairs of additions and subtractions using mental methods.

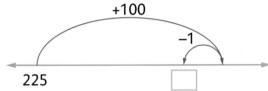

| a | 68 + 31 | b | 58 − 31 | c | 88 − 32 | d | 51 + 42 | e | 63 + 41 | f | 63 + 39 |
| | 68 + 29 | | 58 − 29 | | 88 − 28 | | 51 − 42 | | 63 − 41 | | 63 − 39 |

2 Write the calculation that each number line shows.

a

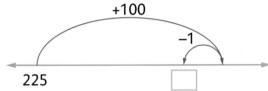

+100
−1
225 ☐

b

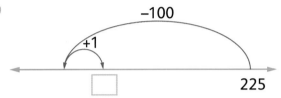

−100
+1
☐ 225

c

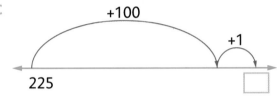

+100
+1
225 ☐

d

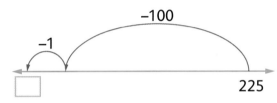

−100
−1
☐ 225

3 Copy these and write the answers.

a 146 + 100
   146 + 102
   146 + 98
   146 + 97

b 246 − 100
   246 − 102
   246 − 98
   246 − 97

c 346 + 203
   346 + 197
   346 − 203
   346 − 197

Try this

Use your skills of specialising and generalising.
Pia uses rounding to help her with a calculation.
She adds a multiple of ten and then subtracts 1.

☐ ☐ ☐ + ☐ ☐ = 195

What could the calculation be?

Find all the solutions. What do you notice?

Working with addition

Explore

How high is this human tower?

Is 100 cm a good **estimate**?

What about 500 cm?

How could you make a good estimate of the total height?

What do you need to think about to help you make the best **prediction** possible?

Maths words

estimate

prediction

decompose

sum

regroup

Learn

You can use addition to work out the total heights.
Estimate: **174 cm + 149 cm**
You can work out the actual total in different ways.

Adding on method

174 + 149 =

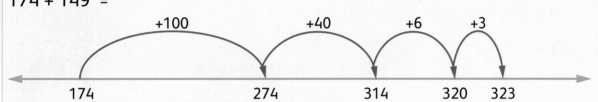

Decomposing method

174 + 149 =
100 + 100 = 200
70 + 40 = 110
4 + 9 = 13 +
           ___
           323

Written column method

100s	10s	1s	Method
1	7	4	4 ones + 9 ones
+ 1	4	9	7 tens + 4 tens, then add one more ten
3	2	3	1 hundred + 1 hundred, then add 1 more hundred
1	1		

How close was your estimate to the actual total? 149 is a near multiple of 10.
How does this help you to think of another way you can add to find the **sum** of 174 and 149?

Practise

1 a Estimate the answers to these.

453 + 230 = ☐   152 + 128 = ☐

176 + 342 = ☐   365 + 278 = ☐

378 + 265 = ☐   559 + 197 = ☐

b Work out the actual answers.

Practise (continued)

2 Use regrouping to help you complete the additions below this example.

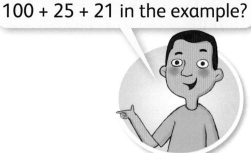

Why do we **regroup** 146 as 100 + 25 + 21 in the example?

375 + 146 = | 375 + 100 + 25 + 21 |

a 450 + 265 = ☐

b 493 + 128 = ☐

Use estimates or a calculator to check your answers.

c 385 + 128 = ☐

d 389 + 234 = ☐

3 There are 465 people at a station. Another 248 people arrive. How many people are at the station in total?

Try this

I know that 356 + 197 = 523 must be wrong because a sensible estimate is 350 + 200 = 550. The answer 523 is too low.

Let's talk

Use estimates like Sanchia's to help you check these additions. Convince a partner that you are correct.

a 309 + 184 = 493

b 478 + 234 = 742

c 249 + 355 = 594

d 686 + 293 = 979

Think about the methods you used for the additions in Practise. Compare them with a partner. Explain why you chose each method. Write some other additions that you would solve using a mental method. Now write some that you would solve using the column method. Use your specialising skills to justify your choices.

Working with subtraction

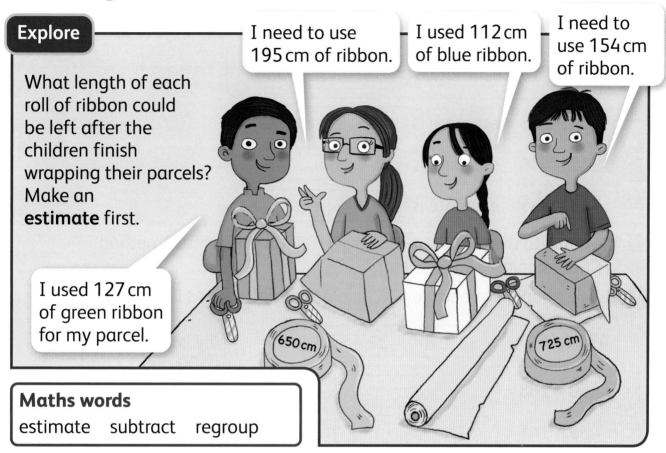

Explore

What length of each roll of ribbon could be left after the children finish wrapping their parcels? Make an **estimate** first.

I used 127 cm of green ribbon for my parcel.

I need to use 195 cm of ribbon.

I used 112 cm of blue ribbon.

I need to use 154 cm of ribbon.

650 cm

725 cm

Maths words
estimate   subtract   regroup

Learn

Estimate: **523 – 195**
Will the actual answer be more or less than your estimate? Let's investigate.

Subtracting back method	Regrouping method	Written column method	

−2 −3 −90 −100
328 330 333 423   523
	(523 = 400 + 110 + 13)
	523 − 195 =
	400 − 100 = 300
	110 − 90 = 20
	  13 − 5 = 8 +
	          328
		100s 10s 1s
		     4̸5̸ 11̸7̸ 13
		 − 1 9 5
		        3 2 8
			Method
			13 ones – 5 ones
			11 tens – 9 tens
			4 hundreds –
			1 hundred

Was your estimate more or less than the actual total?
195 is a near multiple of 100.
How does this help you to think of another way to solve 523 – 195?

Practise

1 a Estimate. Will the actual answer be more or less than your estimate?

725 – 154 estimate is [700 – 150 = 550]

654 – 248 estimate is [   ]

918 – 436 estimate is [   ]

743 – 355 estimate is [   ]

b Complete the subtractions using a mental or a written method. Check against your estimates.

2 Use regrouping to help you complete the subtractions below.

432 – 164 = [432 – 100 – 32 – 32]

a 450 – 268 = [   ]

b 439 – 145 = [   ]

c 523 – 238 = [   ]

d 635 – 270 = [   ]

Why is 164 regrouped as 100 and 32 and 32 in the example?

Use estimates or a calculator to check your answers.

3 There are 834 people at a museum. 349 people go home. How many people are left at the museum?

Let's talk

Elok uses a counting up method to complete the subtraction **624 – 587**.

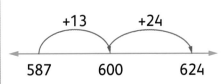

+13   +24

587   600   624

Conjecture why Elok could have used this method instead of others. Make up other calculations to solve using the same method. Explain your choices to convince a partner.

Multiplication and division facts

Explore

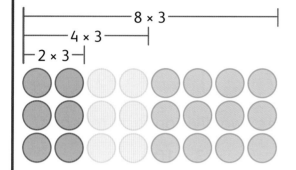

Use your skill of generalising.

Maths words
division
double
triple
multiplication

How can 2 × 3 help you to work out 4 × 3 and 8 × 3?
What relationships can you see?
If we know that 4 × 3 = 12, what two **division** facts can we make?

Learn

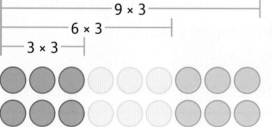

6 × 3 is **double** 3 × 3. What about 9 × 3?
9 × 3 is **triple** 3 × 3.
We can use **multiplication** facts to help us find division facts.
We know that 6 × 3 = 18, so 18 ÷ 3 = 6 and 18 ÷ 6 = 3.

Practise

1 Draw or make an array to show the following facts.
Use the diagrams in **Explore** and Learn to help you.

a 2 × 5, 4 × 5 and 8 × 5

b 2 × 7, 4 × 7 and 8 × 7

c 3 × 2, 6 × 2 and 9 × 2

d 3 × 4, 6 × 4 and 9 × 4

What division facts can you see in your diagrams?

Practise (continued)

2 Complete these sets of facts.

a 2 × 6 = ⬡   b 18 = 2 × ⬡   c 3 × 7 = ⬡   d 15 = 3 × ⬡

4 × 6 = ⬡   36 = 4 × ⬡   6 × 7 = ⬡   30 = 6 × ⬡

8 × 6 = ⬡   72 = 8 × ⬡   9 × 7 = ⬡   45 = 9 × ⬡

3 Complete the multiplication facts. Then write two division facts for each.

a 6 × 10 = ⬡   b 9 × 10 = ⬡   c 3 × 10 = ⬡

d 2 × 8 = ⬡   e 4 × 8 = ⬡   f 8 × 8 = ⬡

4 Write a division fact to solve each problem.
Calculate how many children are in each group.

a At the first party, 27 children sit in nine equal groups.

b Three children leave the first party. The rest sit in four equal groups.

c At a second party, 48 children sit in eight equal groups.

d At a third party, 30 children sit in six equal groups.

e Nine children leave the third party. The rest sit in three equal groups.

Try this

Use your skill of specialising.

9   6   48   4   36   3   24   8

Choose three numbers. Use them to make two multiplication sentences.
Use the same three numbers to make two division sentences.
How many different sets of three numbers can you use in this way?

Quiz

1 Round 16 705 as follows.
  a To the nearest 10   **b** To the nearest 100
  c To the nearest 1 000   **d** To the nearest 10 000

2 Regroup 473 into the number of parts in each question.

  a 473 = ⬭ + ⬭

  b 473 = ⬭ + ⬭ + ⬭

  c 473 = ⬭ + ⬭ + ⬭ + ⬭

3 Copy and complete.

  a 145 + 49 = ⬭   **b** 145 + 99 = ⬭   **c** 145 + 102 = ⬭

  d 175 − 49 = ⬭   **e** 175 − 99 = ⬭   **f** 175 − 102 = ⬭

4 a Match each addition with the correct estimate.

Addition	Estimate
382 + 195	300 + 150 = 450
312 + 148	450 + 250 = 700
438 + 249	400 + 200 = 600

  b Copy and complete the additions. Use a mental or a written method.
  Check your answers against the estimates.

5 Complete. Use a method of your choice.
  a 625 − 246 = ⬭   **b** 625 − 197 = ⬭

6 Complete these, then write two division facts for each one.
  a 2 × 7 = ⬭   **b** 3 × 6 = ⬭

  4 × 7 = ⬭   6 × 6 = ⬭

  8 × 7 = ⬭   9 × 6 = ⬭

12- and 24-hour clocks

0 minutes	1 minute	2 minutes	3 minutes
0 seconds	60 seconds	120 seconds	180 seconds

There are 60 seconds in one minute.

Use your skill of critiquing.

How many minutes is 120 seconds?

How many seconds in 3 minutes?

How many seconds in half a minute?

How many minutes is 90 seconds? And 100 seconds?

Draw your own double number line like this to show minutes and seconds. Continue to 10 minutes.

Study the double number line below for minutes and hours.
Ask your own questions and **convert** (change) between minutes and hours.

0 minutes	60 minute	120 minutes	180 minutes
0 minutes	1 hour	2 hours	3 hours

Maths words

12-hour	a.m.	p.m.	24-hour

Learn

Each day has 24 hours, but clocks only show a **12-hour** day.
Therefore, the hour hand goes around the clock **twice** each day.
To be exact when telling the time, we use **a.m.** for times between midnight and midday. We use **p.m.** for times between midday and midnight.

The diagram shows a **24-hour** timeline divided into these periods.

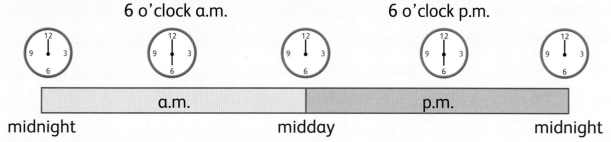

6 o'clock a.m.   6 o'clock p.m.

a.m.   p.m.

midnight   midday   midnight

Point to where these times would go on the timeline above:
- 9 o'clock a.m. and 9 o'clock p.m.
- Half-past ten a.m. and half-past ten p.m.   • 4:45 a.m. and 4:45 p.m.

Practise

1 Read and record these times. Draw clock faces.
   a Quarter past three   b Seven o'clock   c 1:25

2 Estimate the time of each activity. Use a.m. or p.m.
   a Wake up   b Go to sleep   c Eat breakfast
   d Go to school   e Playtime   f Return from school

Try this

Elok has a digital watch.
Guss has a 24-hour watch.

Our watches are correct.
They both show the same time!

Do you agree?
Explain, using your characterising skills.

Learn

There is another way to be exact when telling the time.
There are 24 hours in a day. The 24-hour clock tells the time from 00:00 (midnight) to 23:59 (1 minute to midnight!)

a.m.	p.m.

midnight		midday		midnight
00:00	06:00	12:00	18:00	00:00

This diagram shows how the hours count past 12 at midday.

10 a.m.	11 a.m.	midday	1 p.m.	2 p.m.
10:00	11:00	12:00	13:00	14:00

Practise

1 Convert these times. Say and write them using the 24-hour clock.

   a 1 p.m., 2 p.m., 3 p.m., _____, until 10 p.m.

   b 9 a.m., quarter past nine in the morning, twenty to 10 in the morning

   c 1:30 p.m., 2:45 p.m., 10:50 p.m., 11:59 p.m.

2 Write each time digitally using the 24-hour clock.

a.m.   p.m.

00:30 and 18:00

a   a.m.   p.m.   b   a.m.   p.m.

c   a.m.   p.m.   d   a.m.   p.m.   e   a.m.   p.m.

Calendars and timetables

Explore

How many days in one year? How many weeks in one year? How many days in one week?

Here is a **calendar** for the year 2023. Find your birthday. Does anyone in the class have a birthday in the same week as you? Which **month** has the most birthdays? Find three more dates that are important to you.

Maths words
calendar
month
timetable

January						
Su	Mo	Tu	We	Th	Fr	Sa
1	2	3	4	5	6	7
8	9	10	11	12	13	14
15	16	17	18	19	20	21
22	23	24	25	26	27	28
29	30	31				

February						
Su	Mo	Tu	We	Th	Fr	Sa
			1	2	3	4
5	6	7	8	9	10	11
12	13	14	15	16	17	18
19	20	21	22	23	24	25
26	27	28				

March						
Su	Mo	Tu	We	Th	Fr	Sa
			1	2	3	4
5	6	7	8	9	10	11
12	13	14	15	16	17	18
19	20	21	22	23	24	25
26	27	28	29	30	31	

April						
Su	Mo	Tu	We	Th	Fr	Sa
						1
2	3	4	5	6	7	8
9	10	11	12	13	14	15
16	17	18	19	20	21	22
23	24	25	26	27	28	29
30						

May						
Su	Mo	Tu	We	Th	Fr	Sa
	1	2	3	4	5	6
7	8	9	10	11	12	13
14	15	16	17	18	19	20
21	22	23	24	25	26	27
28	29	30	31			

June						
Su	Mo	Tu	We	Th	Fr	Sa
				1	2	3
4	5	6	7	8	9	10
11	12	13	14	15	16	17
18	19	20	21	22	23	24
25	26	27	28	29	30	

Juy						
Su	Mo	Tu	We	Th	Fr	Sa
						1
2	3	4	5	6	7	8
9	10	11	12	13	14	15
16	17	18	19	20	21	22
23	24	25	26	27	28	29
30	31					

August						
Su	Mo	Tu	We	Th	Fr	Sa
		1	2	3	4	5
6	7	8	9	10	11	12
13	14	15	16	17	18	19
20	21	22	23	24	25	26
27	28	29	30	31		

September						
Su	Mo	Tu	We	Th	Fr	Sa
					1	2
3	4	5	6	7	8	9
10	11	12	13	14	15	16
17	18	19	20	21	22	23
24	25	26	27	28	29	30

October						
Su	Mo	Tu	We	Th	Fr	Sa
1	2	3	4	5	6	7
8	9	10	11	12	13	14
15	16	17	18	19	20	21
22	23	24	25	26	27	28
29	30	31				

November						
Su	Mo	Tu	We	Th	Fr	Sa
			1	2	3	4
5	6	7	8	9	10	11
12	13	14	15	16	17	18
19	20	21	22	23	24	25
26	27	28	29	30		

December						
Su	Mo	Tu	We	Th	Fr	Sa
					1	2
3	4	5	6	7	8	9
10	11	12	13	14	15	16
17	18	19	20	21	22	23
24	25	26	27	28	29	30
31						

Learn

Use your skill of **characterising**.
Bus and train **timetables** use the 24-hour clock.
Why do you think they use this clock and not the 12-hour clock?

Here are some of the stops on the Green bus and the Blue bus routes.
What time does the Green bus leave Elmside station?
Give the time using the 12-hour clock.

Route	Crossley station	Bridge Creek stop	Oak Village stop	Ash Town stop	Elmside station	Pebble Beach stop
Green bus	12:55	__:__	13:25	13:46	14:25	15:09
Blue bus	13:15	13:36	13:58	14:36	15:02	15:35

Practise

1 Use the bus timetable in the Learn box to answer these questions.
  a What time does the Blue bus leave Bridge Creek?
  b Banko arrived at Oak Village at 1:30 p.m. What is the next bus he can catch?
  c Elok takes the quarter past 1 p.m. bus from Crossley station.
    She wants to be at Pebble Beach by 3:45 p.m. Will she arrive on time?
  d Jin's clock shows this time when he arrives at Elmside station.

What time did Jin leave Crossley?

Remember to use the 24-hour clock.

2 Write three story problems about the bus timetable:
  • One that is easy
  • One that is medium
  • One that is difficult.
  Each question must include using either the 12-hour clock or the
  24-hour clock. Display your problems for your classmates to solve.

Let's talk

Use your skill of characterising.

Make a timetable to show your school day.
Use the 24-hour clock to show the times of lessons, breaks, and the start and end of the day.

Quiz

1 Write each time using the 24-hour clock.

a

b

c

☐ p.m.   ☐ p.m.   ☐ p.m.

2 What activities can you do at 1:45 p.m.?

	Pool	Sports court	Gym
10:30–12:00	Free swim	Basketball	Lessons
13:15–14:15	Lessons	Badminton	Exercise class
14:30–16:00	Lane swimming	5-a-side soccer	Free use

Collecting and sorting data

Explore

We can use charts such as a **Venn diagram** or a **Carroll diagram** to **compare** sports.

Maths words

Venn diagram	Carroll diagram
compare	sort

Team sports

Sports using a round ball

	Team sports	Not team sports
Uses a round ball	D	E
Does not use a round ball	F	G

Use your skills of critiquing and classifying to think of a sport to **sort** into each section.

Learn

Pia has drawn this Venn diagram to compare the characteristics of birds and horses.

Birds   Horses

A   B   C

Where would **two eyes** go in the Venn diagram?

Use your critiquing and classifying skills to think of other characteristics to go in each section.

Which animals are we comparing in the Venn diagram?

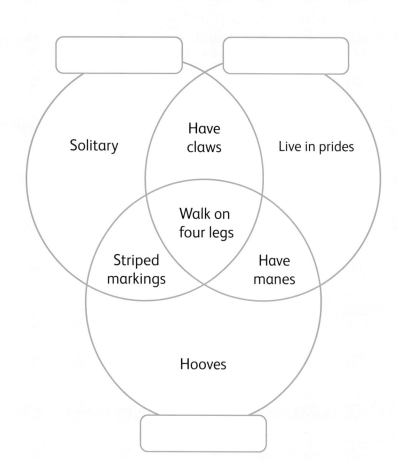

Solitary   Have claws   Live in prides

Walk on four legs

Striped markings   Have manes

Hooves

Practise

Work with a partner or in a group to practise the skills of critiquing, classifying and improving.

1 Choose two or three animals to compare.
Decide how to collect the information.

2 Collect information about the characteristics of your chosen animals.
It may be a good idea to split the task between you, then share your findings with the group.

3 Draw a Venn diagram to present the information.
Work together to discuss which characteristics to place in each section.

Try this

Place the numbers 1 to 30 in diagrams like these.
What do you notice about the numbers that go in sections B and D?

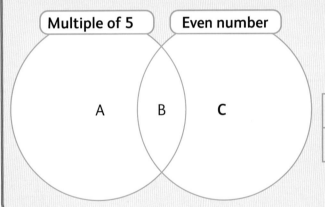

	Multiple of 5	Not a multiple of 5
Even number	D	E
Not even	F	G

Let's talk

Present your Venn diagram from Practise to the class.
Use it to explain the main similarities or differences between the animals.
Look at the Venn diagrams made by other groups.
Ask questions about the data.

Collecting and comparing information

Explore

Each class collects **information** about how they travel to school.

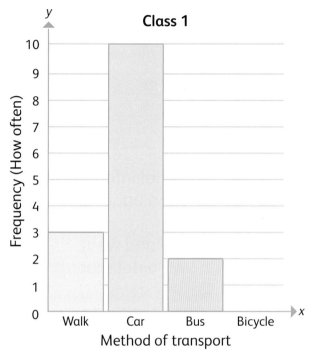

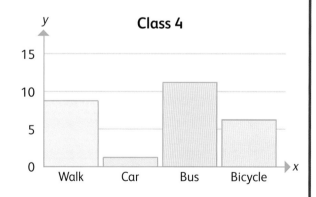

Class 3	
Walk	ℍℍ I
Car	ℍℍ
Bus	IIII
Bicycle	III

Name the chart or table used by each class.
Which charts do you think show the information most clearly?
Which class has the most students? Compare the results for each class. Try to explain the differences in the ways they travel to school. What factors might affect the data? Compare the different charts. What do you notice about the scales used for the **bar charts**?

Maths words

information
bar chart
dot plot

Learn

Sanchia collects information about how many pets people have.
She uses a **dot plot** to record the information.

Number of pets

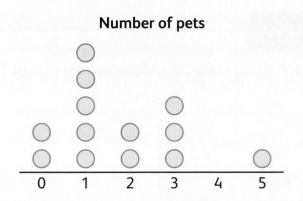

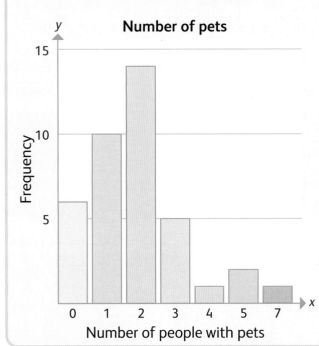

Pia collects the final information and presents it in a bar chart.

What is the same and what is different about a dot plot and a bar chart?
Which would you choose to record information?
When would you choose to use a bar chart to present information?

Practise

1 Investigate the level of difficulty of some class reading books.

2 Discuss which data would be important to collect.
   a Word length
   b Sentence length
   c Number of pages
   d Number of words on a page

3 Decide which data you will collect.

4 Choose one book to investigate.
   Decide how you will collect the data accurately. Explain your choice.
   Record the data carefully in your notebook.

Practise

5 Present your data using a bar chart.
Decide carefully which scale you will use for the frequency.

⭐ 6 Compare your bar chart with those of others.
Interpret the data to compare the difficulty of each book.

Try this ⭐

Use the skill of improving. How accurate were the results of your class reading book investigation? What other factors might affect the level of difficulty? Could fewer words on a page make a book more difficult than more words?

Let's talk

Are there any questions you would like to investigate? What might make it difficult to collect this information?

Quiz

1 Use a Venn diagram like this.

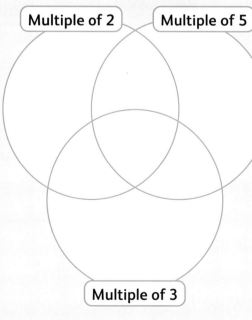

Decide where to write these numbers in the Venn diagram.

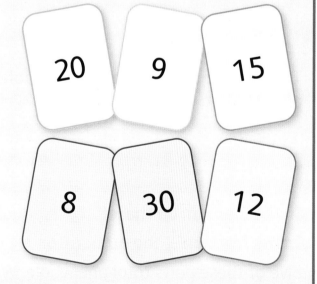

2 When would you choose to use a dot plot to collect or present information?

Parts and wholes

Explore

Jin and Banko are cutting squares to use for their art project.
Jin takes a square and cuts it into two **equal** parts.
Banko takes a square and cuts it into four equal **parts**.

Who has more equal parts?
Whose equal parts are larger?

Maths words

equal
part
whole
fraction
numerator
denominator

Learn

The more equal parts the same **whole** is divided into, the smaller the size of each part.

1				
$\frac{1}{5}$	$\frac{1}{5}$	$\frac{1}{5}$	$\frac{1}{5}$	$\frac{1}{5}$

1		
$\frac{1}{3}$	$\frac{1}{3}$	$\frac{1}{3}$

In the diagram on the left, each equal part is one fifth ($\frac{1}{5}$).

In the **fraction** $\frac{1}{5}$, the **numerator** is 1 and the **denominator** is 5. Why?

Five of these parts total one whole. We can write $1 = \frac{5}{5}$.

What can you say about the diagram on the right?

Practise

1   Work with a partner. Take a sheet of squared paper and carefully cut six strips. Make sure each strip has the same number of blocks as you see here.

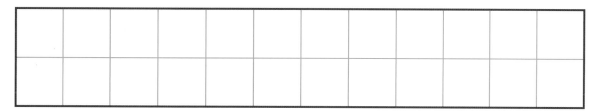

a   Use two strips at a time. Cut them to make the following fractions. Label each part.
   • Halves and quarters   • Thirds and quarters   • Sixths and twelfths

b   Write what you have noticed.
   You can use these words to help you:

   whole   parts   equal   fewer   more   bigger   smaller

2   Use your paper strips to complete these.

   a $\dfrac{6}{\Box} = 1$ b $1 = \dfrac{\Box}{3}$ c $\dfrac{2}{2} = \bigcirc$ d $\dfrac{\Box}{\Box} = 1$

3   Which shape in each pair has the larger fraction shaded? Write the fractions. What do you notice?

a

b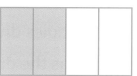

c

Equal shares

Explore

Use the skills of conjecturing
and specialising.
Share one slice of toast equally among
different numbers of children.

Name the **fractions** each time.
What do you notice?

Maths words

fraction	whole	equal
numerator	divide	

Learn

What happens when we share one slice of toast equally among four children?

$\frac{1}{4}$	$\frac{1}{4}$
$\frac{1}{4}$	$\frac{1}{4}$

We can show one **whole** divided into four **equal** parts as the division $1 \div 4$. We represent it as $\frac{1}{4}$.

$\frac{1}{4}$ is a unit fraction because the **numerator** is 1.
Each child gets $\frac{1}{4}$ of a slice of toast.

What happens when we share three slices of toast equally among four children?

$\frac{1}{4}$	$\frac{1}{4}$
$\frac{1}{4}$	$\frac{1}{4}$

$\frac{1}{4}$	$\frac{1}{4}$
$\frac{1}{4}$	$\frac{1}{4}$

$\frac{1}{4}$	$\frac{1}{4}$
$\frac{1}{4}$	$\frac{1}{4}$

We can show three wholes **divided** into four equal parts as the division $3 \div 4$. We represent it as $\frac{3}{4}$.
Each child gets $\frac{3}{4}$ of a slice of toast.
They can get $\frac{1}{4}$ from each slice.

Practise

1 Pia shares one bag of rice equally between bowls.
What fraction of the bag is in each bowl when Pia shares it as follows?

a In three bowls
b In five bowls
c In eight bowls
d In ten bowls

Write a division sentence to match each example.

2 Elok divides three litres of water equally between four jugs.
What fraction of one litre of water is in each jug?

3 Sanchia shares a watermelon equally into six pieces.
She shares an apple equally into two pieces.

a Write fractions to show the size of each piece of fruit.
b Copy and complete the bar models to show the fractions.

Watermelon	Apple

c Which fraction of the whole is larger? Explain why.

Try this

Mum shares three cakes equally between four plates.
She shares a pineapple equally between eight plates.
Next, Mum shares a pie equally across some plates.
One plate of pineapple represents the smallest fraction of a whole.
One plate of cake represents the largest fraction of a whole.

> Among how many plates could
> Mum share the pie equally?
> What could the fraction be?
> **Convince** your partner.

Fractions of shapes and quantities

Explore

Use your skill of convincing. Can you see quarters?
What other **fractions** can you see?

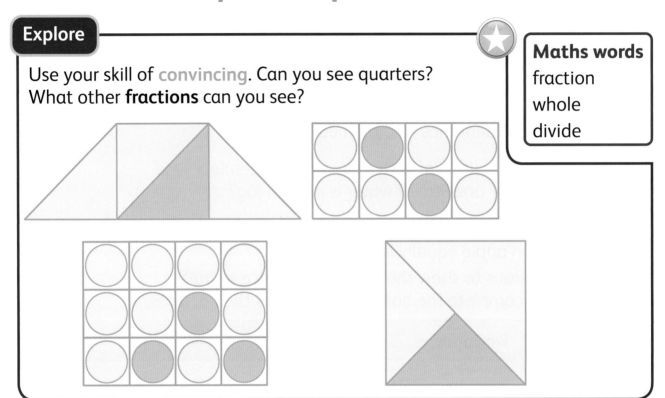

Maths words

fraction

whole

divide

Learn

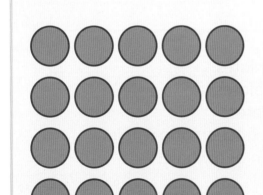

The **whole** is 20 counters. We can use what we know about division to find different fractions of the number of counters.

Remember that $\frac{1}{2}$ represents one whole **divided** by two. What is $\frac{1}{2}$ of 20?

What other fractions of 20 can you find? How does the array help you?

We can also use bar models to represent fractions of amounts.

20				
4	4	4	4	4

We can write:
$\frac{1}{5}$ of 20 = 4

Practise

1 Work with a partner. You need 24 counters.

Find different fractions of 24. Record them as $\frac{\square}{\square}$ of 24 = $\bigcirc$

What do you notice about the different fractions you find?

2 Find the missing values. Record each example as $\frac{\square}{\square}$ of $\bigcirc$ = ?

a

15		
?	?	?

b

28			
?	?	?	?

c

30					
?	?	?	?	?	?

Try this

Use your skill of specialising. Choose three of the digits 1, 2, 3 or 4 each time to make up different fractions of amounts.

$$\frac{1}{\square} \text{ of } \boxed{}\boxed{} = ?$$

How many different answers can you find?

Let's talk

You will need squared paper. Find the following fractions of different rectangles.

$\frac{1}{3}$ $\frac{1}{6}$ $\frac{1}{8}$ $\frac{1}{10}$

Show each fraction on three different rectangles.

Think about the number of squares you will use for each rectangle. This will help you to show the fractions accurately.

Equivalent fractions

Explore

Look at this fraction wall.

What is the same and what is different about each row?

Are any rows missing? What would they look like?

How many tenths are equal to a half?

How many thirds together are greater than a half?

What else do you notice?

Maths word
equivalent

1 whole											
$\frac{1}{2}$						$\frac{1}{2}$					
$\frac{1}{3}$				$\frac{1}{3}$				$\frac{1}{3}$			
$\frac{1}{4}$			$\frac{1}{4}$			$\frac{1}{4}$			$\frac{1}{4}$		
$\frac{1}{5}$		$\frac{1}{5}$		$\frac{1}{5}$		$\frac{1}{5}$		$\frac{1}{5}$			
$\frac{1}{6}$		$\frac{1}{6}$		$\frac{1}{6}$		$\frac{1}{6}$		$\frac{1}{6}$		$\frac{1}{6}$	
$\frac{1}{8}$	$\frac{1}{8}$	$\frac{1}{8}$	$\frac{1}{8}$	$\frac{1}{8}$	$\frac{1}{8}$	$\frac{1}{8}$	$\frac{1}{8}$				
$\frac{1}{10}$	$\frac{1}{10}$	$\frac{1}{10}$	$\frac{1}{10}$	$\frac{1}{10}$	$\frac{1}{10}$	$\frac{1}{10}$	$\frac{1}{10}$	$\frac{1}{10}$	$\frac{1}{10}$		
$\frac{1}{12}$	$\frac{1}{12}$	$\frac{1}{12}$	$\frac{1}{12}$	$\frac{1}{12}$	$\frac{1}{12}$	$\frac{1}{12}$	$\frac{1}{12}$	$\frac{1}{12}$	$\frac{1}{12}$	$\frac{1}{12}$	$\frac{1}{12}$

Learn

$\frac{2}{4}$ is the same as $\frac{1}{2}$.
These are **equivalent** fractions.

$\frac{3}{4}$ is greater than $\frac{6}{8}$.

Do you agree with Sanchia?
Critique the fraction wall.
How can you use it to help you decide?
Do you agree with Guss? Convince your teacher.

Practise

1 Use the fraction wall in **Explore** to write all the equivalent fractions of $\frac{1}{2}$.

2 Which of these fractions have the most equivalent fractions on the fraction wall? Make an estimate for each. Then list all the equivalent fractions. Were your estimates correct?

$\frac{3}{4}$ $\frac{4}{6}$ $\frac{2}{8}$ $\frac{1}{3}$ $\frac{4}{12}$

3 No fractions on the fraction wall in **Explore** are equivalent to $\frac{5}{8}$.

Can you think of any fractions that are equivalent to $\frac{5}{8}$? Name them.

Now, name some fractions that are equivalent to $\frac{5}{12}$.

Try this

Complete the sentences so that each friend gets the same fraction of a whole fruit. **Convince** a partner how you know.

Three oranges shared between four friends is equivalent to ⬜ oranges shared between eight friends.

Four apples shared between ⬜ friends is equivalent to two apples shared between three friends.

⬜ pineapples shared between ten friends is equivalent to two pineapples shared between five friends.

Let's talk

Work with a partner. Take turns to make a shape using two colours of cubes. Say the fractions that are represented. For example:

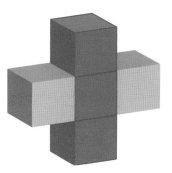

$\frac{2}{5}$ of the shape is orange.

$\frac{3}{5}$ of the shape is purple.

Quiz

1 There are two equal lengths of string.
   Pia cuts the red string into quarters.
   She cuts the blue string into fifths.
   Will a piece of red string be longer
   or shorter than a blue piece?

2 Copy and complete these.

   a $1 = \dfrac{10}{\square}$ **b** $1 = \dfrac{\square}{12}$ **c** $\dfrac{8}{8} = \square$

3 What fraction will each person get?
   a Two children share one sandwich equally between them.
   b Five children share one watermelon equally.
   c Gran shares three pies equally between four children.

4 Which is the bigger amount each time?

   a $\dfrac{1}{2}$ of 12 or $\dfrac{1}{4}$ of 28   **b** $\dfrac{1}{3}$ of 30 or $\dfrac{1}{5}$ of 30

5 What fraction of the shape is shaded for each colour?

6 Write three fractions that are equivalent to $\dfrac{1}{3}$.

7 In Jin's box of pencils, $\dfrac{1}{2}$ of the pencils are red and $\dfrac{2}{4}$ are blue.
   Are there any other colours of pencils in the box?
   Explain how you know.

Units 1–6

1 Which number is larger each time?
   a −5 or 5
   b 0 or −6
   c −10 or −4

2 How many hundreds are in each of these numbers?
   a 4 325
   b 16 075

3 Draw a rectangle that is 4 cm wide and 5 cm long.

4 Take the rectangle you drew in question 3 and do the following.
   a Calculate the area.
   b Calculate the perimeter.

5 Round each number.

	To the nearest 1 000	To the nearest 100	To the nearest 10
3 449			
15 652			

6 Copy and complete.

   a 345 + 101 = ⬭

      345 + 99 = ⬭

   b 345 − 102 = ⬭

      345 − 98 = ⬭

What method will you use to help you solve these addition and subtraction calculations?

7   Sort these times in order from earliest to latest.

17:00   5:30 a.m.   midday   07:50   19:45   7:45 a.m.

8   Draw a Venn diagram to compare different modes of transport.

9   Look at the rectangle and read the statements below.
    Are they true or false? Correct any that are false.
    You could make sketches to help you.

a $\frac{1}{5}$ of the rectangle is a larger part than $\frac{1}{6}$.

b $\frac{2}{5}$ of this rectangle is an equal part to $\frac{3}{10}$.

c $\frac{1}{9}$ of the rectangle is a larger part than $\frac{1}{8}$.

10 Elok divides 3 kg of flour equally between four bowls.
    What fraction of a kilogram is in each bowl?

Flour

3 kg

Calculation

Missing number problems

Explore

Elok and Guss are at a museum shop.
They are thinking about the different items they can buy.

What is the cost of:
- one badge?
- one book?
- one pen?
- one cap?

Maths word
symbol

Learn

At the museum shop, Pia buys a flag and a mug for $8.
We can show this as a picture:

 +  = $8

We can use any **symbols** to represent the unknown values.
For example, we can use ⬜ for the flag and ⭕ for the mug.
The missing number sentence is ⬜ + ⭕ = $8.
Banko buys one flag and gets $7 change from a $10 note.

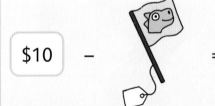

$10 − = $7

Using the symbol for the flag (⬜), we can write the missing number
sentence $10 − ⬜ = $7.
What is the price of the flag? What is the price of the mug?

Practise

1 Work out the unknown price each time.

a

 +  = $10

b

 +  = $18

2 Find the unknown prices. Use both calculations to help you. Write a missing number sentence for each problem. What symbols will you use?

a  +  = $11 $5 −  = $3

b + = 100c   − = 10c

c  +  = $15 $20 −  = $9

3 Find the unknown values.

a ☆ + △ = 100   80 − ☆ = 50

b ▢ + ▢ = 20   ▢ − 7 = ◯

c ◇ + 20 = ▱   70 + ▱ = 100

A blue ribbon and a red ribbon have a total length of 50 cm.
Use symbols to represent the problem.

What are the possible lengths of each ribbon?
Find at least six different solutions.

Addition and subtraction

Explore

Look at the number lines. Some numbers are smudged.

What are the missing numbers?
What are the two calculations?
Talk about any similarities and differences between
the calculations.

Maths words
regroup
estimate

Learn

Here are some methods for solving an addition and a subtraction.
How does each method work? Can you see any **regrouping**?
Which methods do you prefer for these calculations? Why?

Adding on method	Using near multiples method
365 + 257	625 – 298

365 + 257 = 365 + 200 + 35 + 22
       = 565 + 35 + 22
       = 600 + 22
       = **622**
	625 – 298 = 625 – 300 + 2
	       = 325 + 2
	       = **327**

Decomposing method	Column method
365 + 257	625 – 298

365 + 257 =
300 + 200 = 500
 60 + 50 = 110
  5 + 7 = 12 +
      622	

Column method table:

	100s	10s	1s
	$^5\cancel{6}$	$^{11}\cancel{2}$	15
–	2	9	8
	3	2	7

What other methods could you have used?
Did you **estimate** the answers?

Practise

1 How are the answers linked in these calculations?

> 258 – 26 and 258 – 126
> 258 – 26 = 232 is 100 more than 258 – 126 = 132

a 347 + 44 and 347 + 244   b 579 – 133 and 579 – 333
c 425 + 362 and 425 + 562   d 752 – 252 and 752 – 552

Practise (*continued*)

2 Answer these additions using a method of your choice. Estimate first.
Will the actual answer be smaller or larger than your estimate?

a 319 + 295 = ☐

b 375 + 237 = ☐

c 349 + 254 = ☐

d 358 + 279 = ☐

> I estimated 350 + 250 = 600 for one addition. Which one? How did I round the numbers?

3 a Estimate to find the incorrect calculations. Write them down.

285 + 368 = 553   734 − 349 = 385

299 + 399 = 698   634 − 592 = 72

b Make sensible estimates for the incorrect calculations.
Complete them and check against your estimates.

4 Look at the items bought each week, and the price list.

a Estimate how much money was spent each week. Then calculate.

b How much change from $5 was there each week?

Week 1	
Week 2	
Week 3	

Price list	
Bottle of juice	175c
Loaf of bread	149c
Bunch of bananas	95c

Let's talk

Use your skills of specialising and generalising. Use the digits 3, 4, 5 and 9 in different positions in these column calculations.

```
   ☐ ☐ 6              ☐ ☐ 6
 + ☐ 7 ☐            − ☐ 7 ☐
 ─────────          ─────────
```

How many 3-digit answers can you find? What is the largest possible 3-digit answer? And the smallest possible 3-digit answer?

Multiplication table of 7

Explore

Maths words

multiplication
division

You have five hoops each. What different scores can you get?

Imagine scoring 14 with four hoops.

Can your friend score 14 with two hoops?

Learn

We can use **multiplication** facts that we know to help us find new facts.
What do these two arrays show?   What do they show when we join them?

   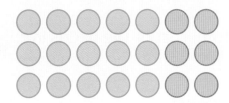

A number of sevens is the total of the same number of fives and twos.

5 × 3 = 15	2 × 3 = 6	7 × 3 = 21
3 × 5 = 15	3 × 2 = 6	3 × 7 = 21

Show that you can use **division**. What is 21 ÷ 7 and 21 ÷ 3?

Practise

1 Complete the multiplication table of seven up to 7 × 10 and 10 × 7.
   Use counters or draw arrays to help you.

2 Look at the example. Then write four facts to match each diagram.

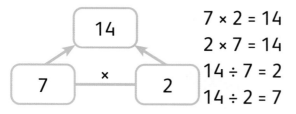

7 × 2 = 14
2 × 7 = 14
14 ÷ 7 = 2
14 ÷ 2 = 7

a

b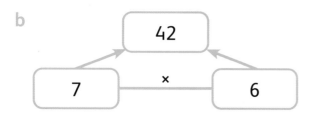

c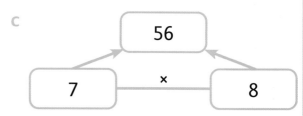

3 a Pens come in packs of twos and fives. Banko buys four packs of each.
     How many pots of seven pens can he make?

  b There are five days in a school week and two days in a weekend.
     How many days in total in six full weeks?

  c A game costs $7. How much do three games cost?

  d Pia shares 35 apples equally among seven baskets.
     How many apples in each basket?

Try this

What is the total cost of the beads in this bracelet?
Write the multiplication facts you use.

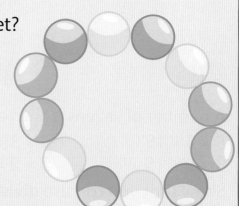

Price list

Yellow beads   2c each

Blue beads   5c each

Green beads   7c each

Multiplying a 2-digit number by a 1-digit number

Explore

What is the same and what is different about these arrays of coins? Use your skill of generalising.

Maths words
multiple
decompose

What is the same and what is different about 4 ones × 3 and 4 tens × 3?
Add another row to each array. What do they show now?
Now try adding another column.

Learn

We can draw open arrays to help us make sense of multiplication.
What calculation does this array represent?

	30	6
4		

Why is 40 × 4 = 160 a sensible estimate?

Will the actual answer be more or less than 160?
We can complete the calculation by finding the total of 30 × 4 and 6 × 4.
30 × 4 is the same as 3 tens × 4, so we can use known facts to help us multiply **multiples** of 10.

Learn (continued)

Here are two more methods to show 36 × 4.
You can **decompose** the multiplication or use the column method.

Decomposing method	Column method

Decomposing method:

$$36 \times 4 \begin{cases} 30 \times 4 = 120 \\ \\ 6 \times 4 = \underline{24} + \\ \quad\quad\quad\quad 144 \end{cases}$$

Column method:

	10s	1s	Method
	3	6	6 ones × 4 = 24 ones
×		4	or 2 tens and 4 ones 3 tens × 4 = 12 tens
	1 4	4	plus 2 more tens = 14 tens
	2		

Can we use doubles to help us find 36 × 4?
How can we use doubles to find 36 × 8?

Practise

1 Use multiplication facts to help you complete these.

a 4 × 6 = ☐
40 × 6 = ☐
4 × 60 = ☐

b 8 × 6 = ☐
80 × 6 = ☐
8 × 60 = ☐

c 3 × 7 = ☐
30 × 7 = ☐
3 × 70 = ☐

d 9 × 7 = ☐
90 × 7 = ☐
9 × 70 = ☐

2 Sketch and complete these open arrays.
Write the matching multiplication and answer.

a
        20   8
3 [   ][   ]

b
        20   8
6 [   ][   ]

c
        30   3
7 [   ][   ]

d
        60   6
7 [   ][   ]

Practise *(continued)*

3 Solve these multiplications using the method shown.

a 37 × 6 = ☐

b 37 × 3 = ☐

c 43 × 4 = ☐

d 43 × 8 = ☐

$$35 \times 7 \begin{cases} 30 \times 7 = 210 \\ \\ 5 \times 7 = \underline{\ 35} + \\ \qquad\qquad \underline{\underline{245}} \end{cases}$$

4 Use the column method to answer these.
Make an estimate first.

a 32 × 6 = ☐

b 45 × 7 = ☐

Remember to check your answers using your estimates and a calculator.

c 63 × 3 = ☐

d 56 × 4 = ☐

Try this

There are 24 hours in one day.

a How many hours are there in three days?

b How many hours are there in a week?

c How many hours are there in nine days?

Multiplying a 3-digit number by a 1-digit number

Explore 🔗 ⭐

Maths words
decompose
multiple

You are playing target games using beanbags.

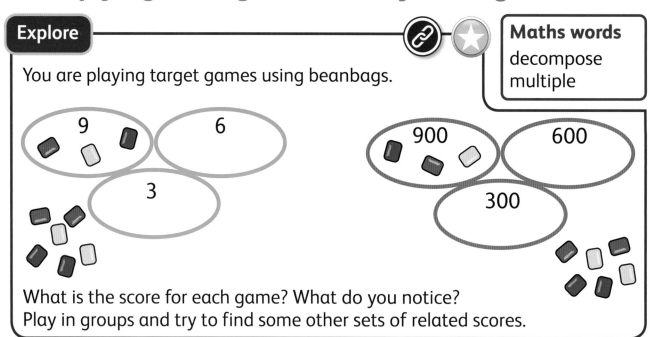

What is the score for each game? What do you notice?
Play in groups and try to find some other sets of related scores.

Learn

What calculation does this array represent?
The 3-digit number has been **decomposed**.

	200	80	2
3			

Why is 300 × 3 = 900 a sensible estimate?
Will the actual answer be more or less than 900?
We can complete the calculation by finding the total of 200 × 3,
80 × 3 and 2 × 3.
200 × 3 is the same as 2 hundreds × 3, so we can use known facts
to help us multiply **multiples** of 100.

Look at this column
method example.
Look at the way it
matches the parts
of the array.

100s	10s	1s	Method
2	8	2	2 ones × 3 = 6
×		3	8 tens × 3 = 24 tens
			or 2 hundreds and 4 tens
8	4	6	2 hundreds × 3 = 6 hundreds plus 2 more
2			hundreds is 8 hundreds

We can check the answer using our estimate or a calculator.

Practise

1 Use multiplication facts to help you complete these.

a $2 \times 9 =$ ☐   b $5 \times 9 =$ ☐   c $7 \times 9 =$ ☐

$200 \times 9 =$ ☐ $500 \times 9 =$ ☐ $700 \times 9 =$ ☐

$2 \times 900 =$ ☐ $5 \times 900 =$ ☐ $7 \times 900 =$ ☐

2 a Complete the arrays. Write the matching multiplications and answers.

	100	50	8
6			

	100	8
9		

b Draw arrays and solve these multiplications. Make an estimate first.

139×4 247×3 117×8

3 Use the column method to answer these. Make an estimate first.

a $123 \times 6 =$ ☐   b $135 \times 7 =$ ☐

c $263 \times 3 =$ ☐   d $246 \times 4 =$ ☐

4 Pia cuts 7 lengths of ribbon. Each is 138 cm long.
Guss cuts 6 lengths of ribbon. Each is 159 cm long.
Who uses a greater length of ribbon in total? How much more?

Let's talk

Use estimates to sort these calculations into a table like the one below.

535×6 458×6 390×7 641×9 234×8 782×7 893×2

Estimate is less than 2 000	Estimate is between 2 000 and 5 000	Estimate is greater than 5 000

Make up another calculation to go in each box.

Quiz

1 Find the value of ☆ and ▢

    a ☆ + ▢ = 100   **b** 84 − ▢ = 42

2 Copy and complete.

    a 365 + 258 = ▢   **b** 358 + 265 = ▢

    c 632 − 287 = ▢   **d** 632 − 596 = ▢

3 There are 7 days in a week. How many days in:

    a 4 weeks?   **b** 7 weeks?

    c 9 weeks?   **d** 10 weeks?

4 True or false? Correct any that are false.

    a $7 \times 40 = 4 \times 70$

    b $5 \times 80 < 4 \times 90$

    c $80 \times 8 > 90 \times 7$

5 Use estimates to check if any of these are wrong.

    a $45 \times 7 = 275$ **b** $8 \times 74 = 592$

    c $63 \times 6 = 378$ **d** $9 \times 43 = 357$

6 Find the missing numbers.

    a ▢ $\times 6 = 2\,400$ **b** $600 \times$ ▢ $= 2\,400$

    c $800 \times 7 =$ ▢   **d** $700 \times$ ▢ $= 5\,600$

7 Copy and complete.

    a $123 \times 4 =$ ▢   **b** $123 \times 8 =$ ▢

    c $157 \times 3 =$ ▢   **d** $157 \times 6 =$ ▢

Certain, impossible, likely, unlikely

Explore

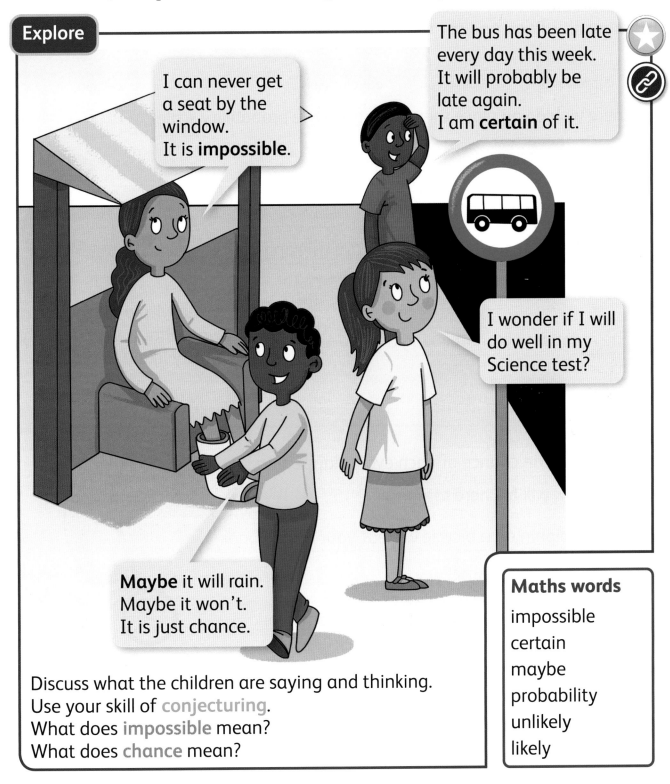

I can never get a seat by the window. It is **impossible**.

The bus has been late every day this week. It will probably be late again. I am **certain** of it.

I wonder if I will do well in my Science test?

Maybe it will rain. Maybe it won't. It is just chance.

Discuss what the children are saying and thinking.
Use your skill of conjecturing.
What does impossible mean?
What does chance mean?

Maths words

impossible
certain
maybe
probability
unlikely
likely

Learn

Probability is how likely it is that an event or something will happen.

This is a probability scale. It includes the words we use to describe probability. Read and say the words. Discuss them with a partner.

| impossible | unlikely | equally likely | likely | certain |

Where would you place these events on the probability scale?

It will rain tomorrow.

The 2D shape you are thinking of has more than two sides.

I will guess the exact number you are thinking of.

You will fly to the moon on a green table.

The number you are thinking of is odd or even.

Agree with a partner on an event that is very **unlikely** or very **likely** to happen.

Practise

1 Describe the chance of each event happening.

a Snow will fall here tomorrow.

b A dolphin on a bicycle will eat your Maths book.

c A book will contain words.

d I will remember to complete my homework.

2 Write an event for the probability words each time.

a It is impossible that ___

b It is very unlikely that ___

c It is likely that ___

d It is certain that ___

Probability experiments

Explore

Maths words
random   chance
experiment

Use your skill of conjecturing.
The children are playing a guessing game.

Banko picks a card at **random**. He does not look as he chooses. What is the **chance** of the shape on his card having an even number of vertices?

Now Jin is going to pick a card at random.
Describe the likelihood of picking a card with a black shape on it.
Is the card he picks more likely to have a triangle or a square on it?
Which child is more likely to pick a circle on their turn?

Learn

Elok and Guss do an **experiment** to test a spinner.
They record their results using a dot plot.

I like this spinner.
It is very likely to
score a 6.

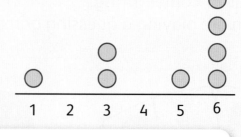

We have only done eight spins.
We need to test it more often.

Practise

1 Use a dot plot to collect results from spinning
a spinner like this 20 times. Compare the results
of your experiment with others in the class.
What do you notice?

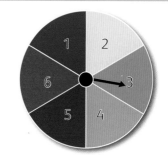

2 Ask your teacher to place some cubes of different
colours in a bag without you seeing.
Take turns to pull out one cube without looking, then put it back.
Record the colours that each of you choose.
After five tests, predict which colour you will choose most often.
Predict again after 10 tests. Predict again after 20 tests.
Write the colours of all the cubes you think are in the bag.
Open the bag to check if your prediction was correct!

3 Play a game of conjecturing and convincing with a partner. Decide who
is Player 1 and Player 2. Each player needs four counters.
Without each other seeing, make a tower of one, two, three or four counters.
Show your towers. If they are the same height, Player 1 wins.
If not, Player 2 wins.
Record the results.
Which player is more likely to win?
Take turns to be each player.

Player 1 wins   Player 2 wins

Try this

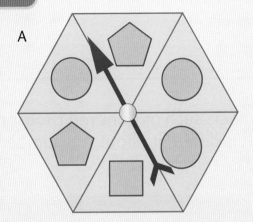

A

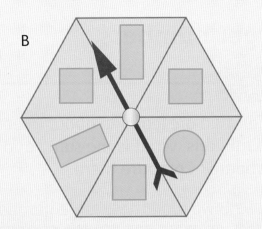

B

Use a dot plot to collect results from spinning spinners like these.
Compare the results after 10 spins each. Complete more tests.
Describe the chance of landing on a circle or a square on each spinner.

Quiz

1 Use the language of probability to describe what will happen in this soccer match.

2 What results could you expect after spinning a spinner like this three times?

What might the results be if you did 200 spins?

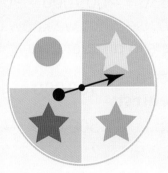

Larger numbers

Explore

Pia and Jin are learning about larger numbers such as **hundred thousands** and **millions**. What is the same and what is different about the numbers on the board? **Compare** them.

How do we read these large numbers?

1 000 000s	100 000s	10 000s	1 000s	100s	10s	1s
		4	9	4	2	8
	4	9	4	2	8	3
4	9	4	2	8	3	6

Maths words

hundred thousands
millions   compare
round

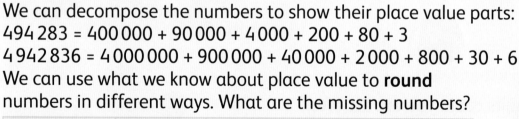

The digit 4 appears twice in each number. Does it have the same value each time?

How many ten thousands does each number have? How many thousands?

Learn

Remember to think about the rules for rounding. Which digit will you check each time?

We can decompose the numbers to show their place value parts:
494 283 = 400 000 + 90 000 + 4 000 + 200 + 80 + 3
4 942 836 = 4 000 000 + 900 000 + 40 000 + 2 000 + 800 + 30 + 6
We can use what we know about place value to **round** numbers in different ways. What are the missing numbers?

Round to the nearest:	100 000	10 000	1 000	100
494 283	500 000	490 000	494 000	494 300
4 942 836	?	?	?	?

Practise

1 Read these numbers. Then write them as numerals.
  a Two hundred and fifty thousand, three hundred
  b Two hundred and fifty-three thousand, three hundred and twenty-two
  c One million, two hundred and fifty thousand, three hundred

2 Decompose these numbers to show their place value parts.

  a 54 203 = ☐ + ☐ + ☐ + ☐ + ☐

  b 254 135 = ☐ + ☐ + ☐ + ☐ + ☐ + ☐

  c 4 375 069 = ☐ + ☐ + ☐ + ☐ + ☐ + ☐ + ☐

3 65 324   273 052   3 246 358

  Round each number above to the nearest:
  a 100 000   b 10 000   c 1 000   d 100   e 10

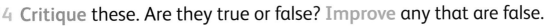

4 Critique these. Are they true or false? Improve any that are false.
  a 135 345 > 135 435   b 326 400 < 3 264 hundreds
  c 232 640 < 2 325 404   d 407 231 < 1 072 319

Try this

Banko is thinking of a number. It rounds to 30 000 to the nearest 10 000.
When he counts in steps of 1 000 from his number, one of the numbers he says
is 35 232. What could the number be? Find all the possibilities. Be convincing.

Let's talk

Use your skills of specialising and generalising. A fact book rounds the
population of some European cities to the nearest 100 000.

City	Population
Berlin	3 700 000
Madrid	3 200 000
Rome	2 800 000
Sofia	1 200 000

Why do you think the numbers were rounded?
What could the actual populations be?
What is the largest possible population?
What is the smallest possible population?

Working with sequences

Guss and Elok arranged some boxes and tins in piles like this.

Maths words	
sequence	term
rule	recursion

 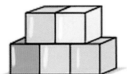 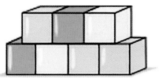 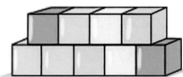

 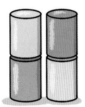 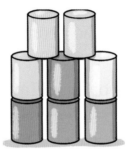 

What patterns can you see?
How many boxes and tins would be in the next few piles?
How many boxes and tins would be in the previous pile?

Learn

The patterns shown in **Explore** are called **sequences**.
The values in a sequence are connected by a rule.
These values make up the **terms** of the sequence.
Count the number of boxes: 3, 5, 7, 9 ___
The term-to-term **rule** for the boxes is:
Next term is previous term add two.

A term-to-term rule is also called a **recursion** rule. Can you think why?

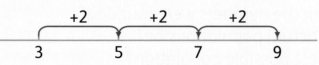

What is the term-to-term rule for the tins?

Practise

1 Find the values of the missing terms.

Term-to-term rule	Sequence
a Next term is previous term subtract 5	22, 17, 12, ☐, ☐, ☐, ☐
b Next term is previous term add 4	−16, ☐, −8, ☐, ☐, 4
c Next term is previous term multiplied by 3	☐, ☐, 9, 27, ☐, ☐
d Next term is previous term divided by 5	☐, 1 000, 200, ☐, ☐

2 Work out the next values in each sequence. Write the term-to-term rule.

11, 31, 51, 71, 91, 111, 131   Rule is: Add 20

a 22, 16, 10, ☐, ☐, ☐   Rule is: _____

b 3, 6, 12, 24, ☐, ☐, ☐   Rule is: _____

c 160, 80, 40, ☐, ☐, ☐   Rule is: _____

d 1, 4, 16, ☐, ☐   Rule is: _____

e 2 654, 3 654, 4 654, ☐, ☐, ☐   Rule is: _____

 3 **Convince** a partner about which sequences will have zero as one of the values. Use your **generalising** skills.

a 25, 30, 35, 40, ___
b 95, 90, 85, 80, ___
c 41, 36, 31, 26, ___
d 88, 86, 84, 82, ___
e 42, 39, 36, 33, ___
f 28, 25, 22, ___
g 321, 311, 301, ___
h 550, 540, 530, ___

Does each sequence increase or decrease?

Guss makes up a sequence of numbers.

The third term in this sequence is 10.

☐ , ☐ , **10** , ☐ , ☐ , ☐

What could the missing terms and the rule be?
Could one of the terms be a negative number, or are they all positive?
Make up at least three possible sequences and write the rule each time.

Learn

This sequence follows the rule: Add 3 and multiply by 2.

☐ , **10** , **26** , **58** , ☐ , ☐ , ☐

What are the missing values in the sequence?
How do you know?

Here is another sequence.
What do you notice about it?

☐ , **184** , **88** , ☐ , **16** , ☐ , ☐

The term-to-term rule is: Divide by 2 and subtract 4.
What are the missing values?

Practise

1 Four different sequences follow the rule: Subtract 2 and multiply by 10.
What is the value of the next term after each number?

a (2)   b (12)   c (9)   d (102)

2 Copy and complete.

	Rule	1st term	3rd term
a	Add 3 and divide by 3	24	
b	Subtract 1 and multiply by 5	3	
c	Double and add 3	10	
d	Halve and subtract 1	34	

Let's talk

A sequence of whole numbers follows the rule: Add 1 and double. Try some different starting numbers for the sequence and write the first five terms.

Use the skills of conjecturing and generalising to answer these questions:

a Do any of your sequences contain only odd numbers? Explain your thinking.

b Which term can be an odd number?

c Can you predict whether or not an odd number will ever appear beyond the first five terms? Give some examples.

Even and odd numbers

The children are finding out more about **even** and **odd** numbers.

Which numbers do the diagrams represent?
Look at the rows.
What do you notice about them?

Maths words
even   odd

What do you notice about the totals when you add pairs of even and odd numbers?

even + even

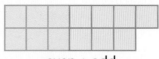

even + odd

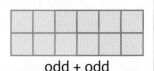

odd + odd

The second diagram shows 6 + 7. We can also write this as 6 + 6 + 1.
Think about why this is. Why do you think the total is odd?
The third diagram shows 5 + 7. We can show why the total is even because:

$(4 + 1) + (6 + 1) = 12$

$4 + \textcircled{1} + 6 + \textcircled{1} = 12$

$4 + 6 + 2 = 12$

Practise

1 Use your generalising skills. Choose numbers from 1 to 10 each time.
   Draw diagrams to show whether these statements are true or false.

   a odd + even = even   b even + even + even = even

   c even + odd + even = odd   d odd + even + odd = odd

   e odd + odd + odd = even   f odd + odd + odd + odd = odd

2 Complete the addition sentences to show why the totals are odd each time.

   a    b

   ( ) + ( ) = ( ) + ( ) + ( )   ( ) + ( ) = ( ) + ( ) + ( )

   c ▭▭▭▭▭▭▭▭▭   ( ) + ( ) = ( ) + ( ) + ( )

3 Write addition sentences to show why the totals are even or odd.

   6 + 3
   6 + (2 + 1)
   6 + 2 + 1
   Total is odd

   a 9 + 5

   b 7 + 6

   c 6 + 5 + 3

Let's talk

Use your skills of classifying and generalising.
Are the answers to these calculations even or odd numbers?
You do not need to work out the answers.

   64 + 22   23 + 64   22 + 65

   90 + 31 + 22   190 + 310 + 221   90 + 22 + 31

Learn

We can explore subtracting even and odd numbers in a similar way.
Think about 10 – 4 and 10 – 3.
The diagram shows the difference between each pair of numbers.

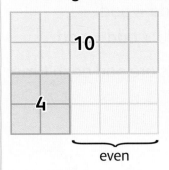

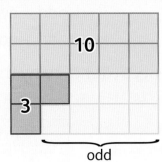

even

odd

What do you notice?
Use the skill
of generalising.

Practise

1 Choose numbers from 1 to 10 and sketch diagrams to
check whether the following statements are true or false.

a odd – even = even

b even – even = even

c even – odd = odd

d odd – odd = odd

e even – odd – odd = even

f odd – even – even = even

2 Complete a table like the one below.
Classify these calculations.
Do not work out the answers.

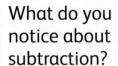

What do you
notice about
subtraction?

63 – 24

144 – 78

87 – 49

299 – 153

346 – 88

268 – 135

Even answer	Odd answer

Try this

Here are two sequences of numbers.

| 10, 8, 6, 4, 2, ___ |

| 1, 3, 5, 7, 9, ___ |

What do you notice about the numbers in each sequence?
What are the rules?

Use what you know about adding or subtracting even and odd numbers to show why this happens. Convince others of your ideas.

The relationship between factors and multiples

Explore

Critique this diagram. Why might it be useful?

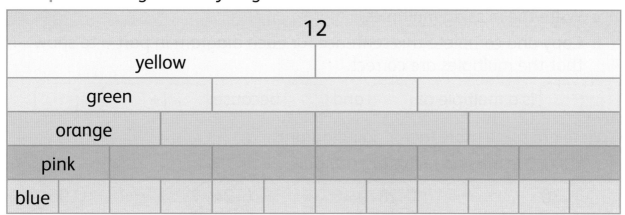

What is the value of a yellow bar? What is the value of each of the other bars?
Use the number 12 each time to make up different multiplication and division sentences.
For example: 6 × 2 = 12 and 12 ÷ 2 = 6.
Use the values of the bars to help you.

Maths words
multiple
factor

Learn

A whole number is a **multiple** of each of its **factors**.
12 is a multiple of 3 and 4 because 12 ÷ 3 = 4 and 12 ÷ 4 = 3.
3 and 4 are factors of 12 because 3 × 4 = 12.
Factors are numbers that divide exactly
into another number with no remainder.

What do you notice about the way the
factors of 12 are joined in this diagram?
We say that these are factor pairs of 12.

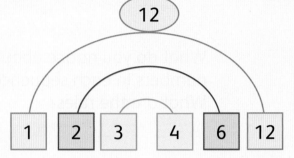

Practise

1 Look at these factor pair diagrams.

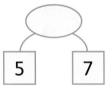 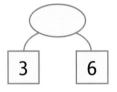 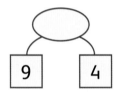 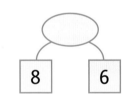

   a Write the missing multiples.

   b Copy and complete this sentence for each diagram in part a to show
     that the multiples are correct.

      ⬚ is a multiple of ⬚ and ⬚ because ⬚ ÷ ⬚ = ⬚

2 Here are some more factor pair diagrams.
   a Write the missing factor in each pair.

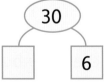 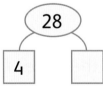 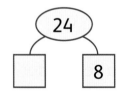 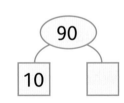

   b Copy and complete this sentence for each diagram in part a to show
     that the factors are correct.

      ⬚ and ⬚ are factors of ⬚ because ⬚ × ⬚ = ⬚

Practise *(continued)*

 3 **Classify** these sets of numbers. Find the odd one out each time. Use factors to help you explain how you know.

a 50, 35, 24, 20, 15

b 20, 100, 50, 35, 40

c 42, 28, 40, 16, 12

d 7, 21, 49, 56, 37

Try this

I made an array to show that 6 and 4 are the only factors of 24.

I made an array to show that 20 is a multiple of 3 and 6.

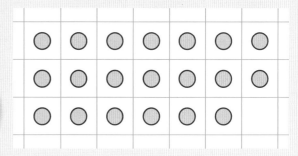

Do you agree with the children? **Critique** their work.
Explain your thinking. Can you **improve** their work in any way?
Show your ideas.

Let's talk

Practise classifying. Where will each number go in the diagram?

(16) (30) (24) (15) (48) (62) (56)

Talk to a partner about how you know.

	Multiple of 8	Not a multiple of 8
Multiple of 6		
Not a multiple of 6		

Think of two other numbers to add to the diagram. Where will you put them? Why?

Quiz

1 Write the missing numbers in each sequence. What is the rule?

a ☐ , 19, 14, 9, ☐ , ☐ , ☐

b ☐ , 6, 12, 24 , ☐ , ☐

2 A sequence has this recursion rule: Add 3 and multiply by 10. The first term is 2. What is the value of the third term?

3 Choose numbers from 1 to 10 to make each statement true.

a ☐ + ☐ = even   b ☐ − ☐ = even

c ☐ − odd = odd   d odd − ☐ = odd

e even + ☐ = odd

4 Write a factor pair for each multiple. Write a multiplication sentence to show why you have chosen the numbers.

a 10   b 18   c 20   d 42

5 Explain why 25 is not a multiple of 3.

Symmetry

Explore

Elok folds a square to find the **lines of symmetry**.

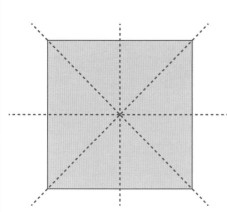

My square has four lines of symmetry.
Two lines are **diagonals**!

Maths words

line of symmetry
diagonal
horizontal
vertical
2D shape
symmetry

Trace and cut out these shapes. Fold them to find the lines of symmetry.
Use your skills of critiquing and improving.
Do any of the shapes have lines of symmetry that are diagonals?

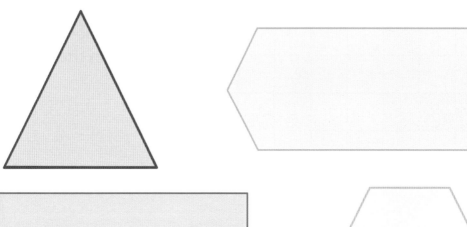

Learn

Shapes can have lines of symmetry that are **horizontal** and **vertical**.

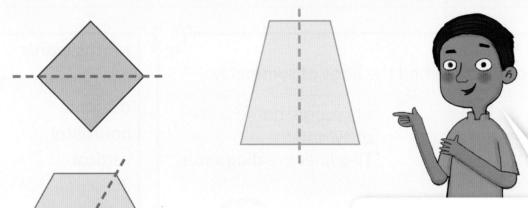

We show lines of symmetry with broken lines, like this.

Patterns can also have lines of symmetry.
Use a mirror to find the lines of symmetry in these patterns.

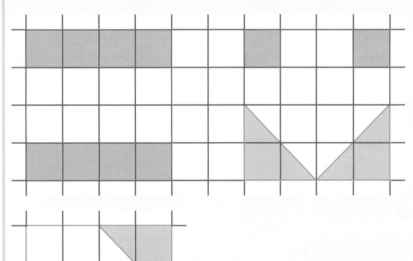

Try to find all the lines of symmetry in these patterns.

Practise

1 Use your skills of **characterising** and **classifying**. Are all these lines of symmetry correct? Discuss and predict. Use a mirror to check. Or, trace the shapes onto paper, cut them out and fold them.

a
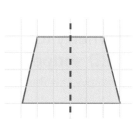

b
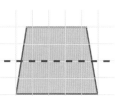

c
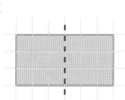

d

e
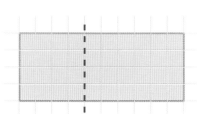

f
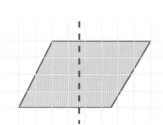

g
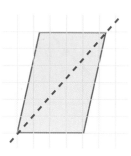

2 How many lines of symmetry does each pattern have?

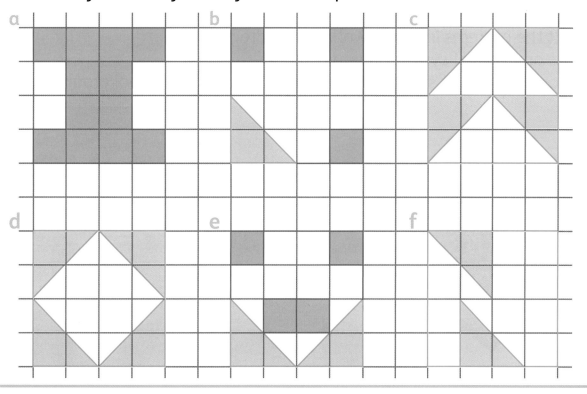

Try this

This regular octagon has at least two lines of symmetry.
Find all the lines of symmetry.

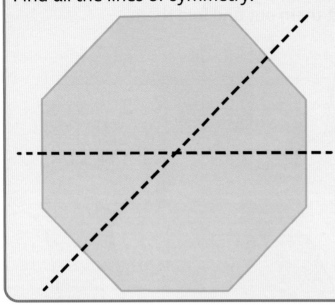

Investigate the number of lines of symmetry for other regular **2D shapes**.

Let's talk

Look at the patterns in the artwork and photograph.
Discuss any lines of symmetry you can find.
Look for examples of **symmetry** in your classroom and school grounds.

3D shapes and nets

Explore

Look at these **3D shapes**. Try to name each shape.
Describe their properties to a partner.

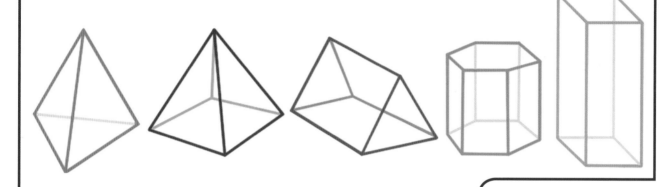

a Which shape has the most triangular **faces**?
b Which shapes have an odd number of faces?
c Which shape has the most rectangular faces?
d How many rectangular faces in total?

Maths words
3D shape
face   net
vertices   edge

Learn

A **net** is a 2D shape that
folds to make a 3D shape.

1   2   3

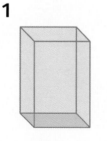 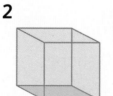 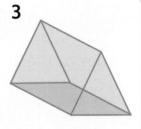

A   B   C

Sanchia unfolds the boxes.
Match the boxes to the nets.
Try unfolding some boxes
or other containers made
from card.

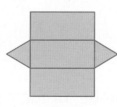 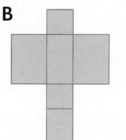 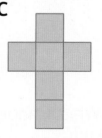

Practise

1 Match the net to the shape. Write the correct letter and number.

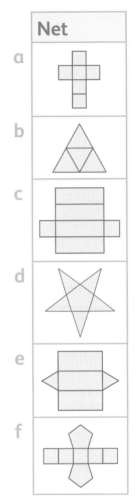

Think about the faces and their shapes.
Will you think about the number of vertices?

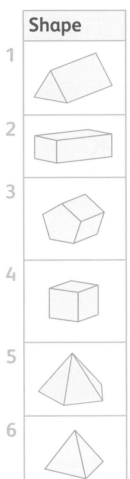

2 Look at these 3D shapes.

  a   b   c   d

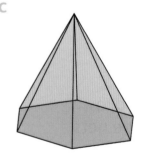

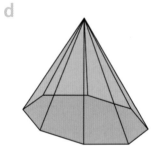

Which shapes have an even number of faces?

Which shapes have an odd number of vertices?

Try this

Each net forms a 3D shape.
Write the number of faces, **vertices** and **edges** for each shape.

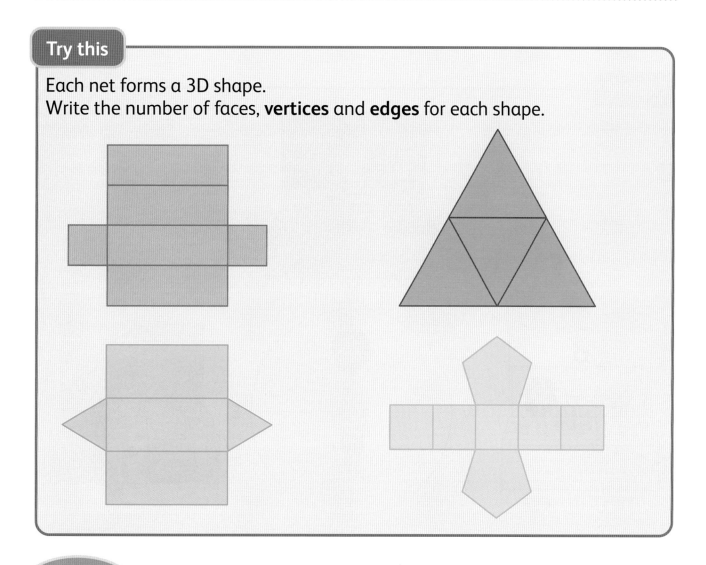

Let's talk

Cubes have only square faces.

Cuboids have only rectangular faces.

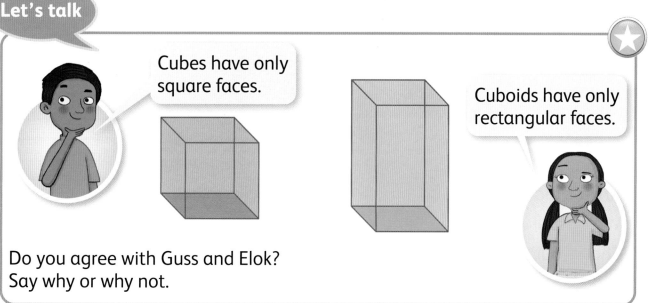

Do you agree with Guss and Elok?
Say why or why not.

Angles and turns

Explore

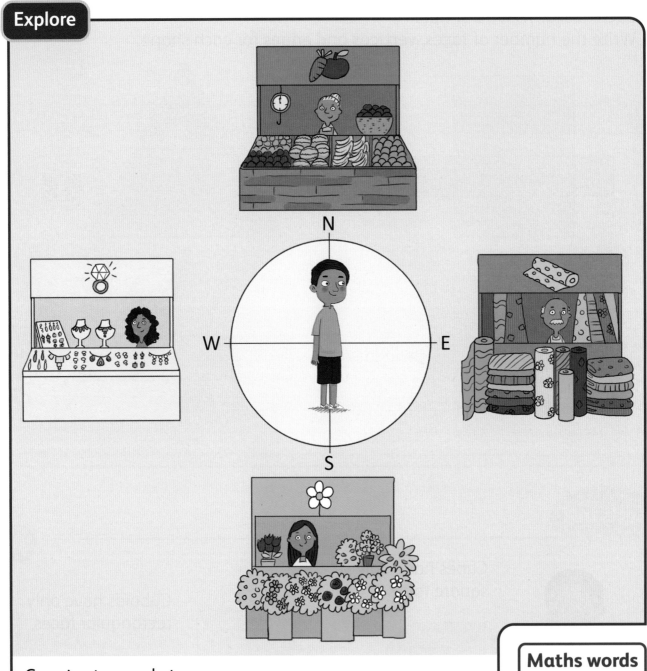

Guss is at a market.
If Guss makes a half **turn**, he will face the jewellery stall.

Which stall will Guss face if he makes a quarter turn?

Share your answer with a partner.
Do you both agree?

Maths words
turn
right angle
degree
acute
obtuse

Learn

A quarter turn is also called a **right angle**.
A right angle measures 90 **degrees**.

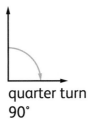

quarter turn
90°

A half turn is 180 degrees.

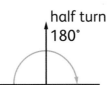

half turn
180°

How many degrees is a whole turn?

whole turn
360°

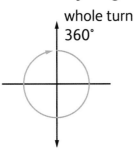

If a turn is less than a right angle, it is called **acute**.

acute

If a turn is between a quarter turn and a half turn, it is called **obtuse**.

obtuse

Practise

1 Decide if each angle shows a quarter turn, an acute angle or an obtuse angle.

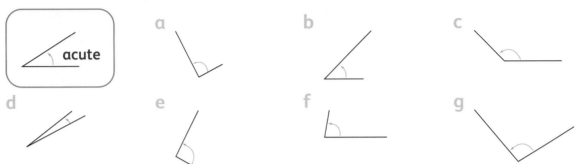

2 a The needle on this dial turns from 3 to 7.
   Name the turn.
   b Is the turn obtuse? How do you know?
   c From 7, the needle moves a quarter turn.
   What number does it point to?

3 Draw these shapes and label the angles.
   a A triangle with three acute angles
   b A triangle with one obtuse angle
   c A four-sided shape with two acute angles and two obtuse angles
   d A hexagon with three obtuse angles and three acute angles

Try this

Find some 2D shapes in your classroom.
Draw them and label each **interior** (inside)
angle as **acute**, **obtuse** or **right angle**.

Let's talk

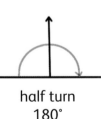

quarter turn
90°

Banko says he can jump a quarter turn, which is 90 degrees.
Jin says he can jump two quarter turns, which is 180 degrees.
Sanchia says she can jump a three-quarter turn.
How many degrees is this?
Make a quarter turn, a half turn and a three-quarter turn.
Say the number of degrees as you turn.

half turn
180°

Quiz

1 Show all the lines of symmetry on shapes like these.

a   b   c

  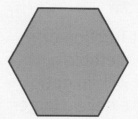

2 Name the 2D faces on these 3D shapes.

a   b   c

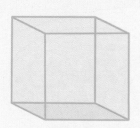 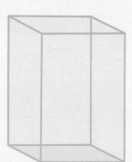 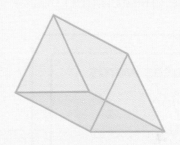

3

Name the 3D shape that this net makes.

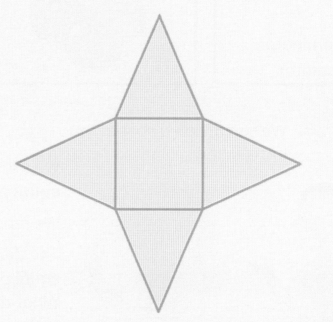

4 a Draw an acute angle.
   b Draw an obtuse angle.

113

Equal parts

Explore

What fraction of the vegetable pizza will Jin get? Will he have a larger or smaller piece of the cheese and tomato pizza? Why? What fraction of each of the other items will the children get?

Maths words

divide
part
whole
equal
denominator
numerator

Let's share each item of food and drink equally!

I don't want any cheese and tomato pizza!

Learn

We can **divide** a shape into eight equal **parts**.

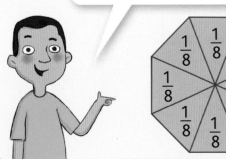

We show one **whole** divided into eight **equal** parts as the division $1 \div 8$.

We represent this as $\frac{1}{8}$.

The **denominator** 8 tells us how many equal parts the whole is divided into. What does the **numerator** 1 tell us?

Practise

1 Use the skill of critiquing. True or false? Improve any that are false.

   a $\frac{1}{3}$ of an orange is larger than $\frac{1}{4}$ of the same orange.

   b $\frac{1}{8}$ of a pineapple is larger than $\frac{1}{6}$ of the same pineapple.

   c One banana cut into ten equal parts will give larger pieces than the same banana cut into nine equal parts.

2 Pia is planting seeds. She divides one whole bag of compost equally between some flower pots. What fraction of the whole bag of compost is in each flower pot when Pia uses:

Compost

   a five flower pots?

   b seven flower pots?

   c nine flower pots?

   d twelve flower pots?

Write a matching division sentence each time: $1 \div \boxed{} = \frac{\boxed{}}{\boxed{}}$

3 Some charities equally share the total money raised from a fun run. They each get $\frac{1}{6}$ of the money. How many charities are there?

Try this

Banko draws these diagrams to help him solve a problem.

1			
$\frac{1}{4}$	$\frac{1}{4}$	$\frac{1}{4}$	$\frac{1}{4}$

1			
$\frac{1}{4}$	$\frac{1}{4}$	$\frac{1}{4}$	$\frac{1}{4}$

1			
$\frac{1}{4}$	$\frac{1}{4}$	$\frac{1}{4}$	$\frac{1}{4}$

What could the problem be?
Make up three different examples.

Finding fractions of shapes and quantities

Explore

This is $\frac{1}{3}$ of a class. What **fraction** is not shown?
How many children are in the class?

Maths words
fraction
divide

Learn

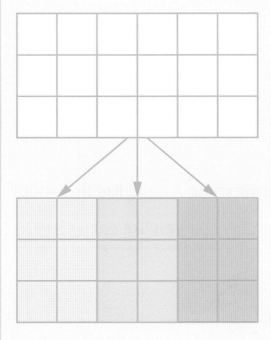

This diagram shows how you can find $\frac{1}{3}$ of a number when you **divide** it by 3.

Describe to a partner how Jin's method works.

$18 \div 3 = 6$

$\frac{1}{3}$ of 18 is 6

What other fractions of 18 can you find using the diagram?

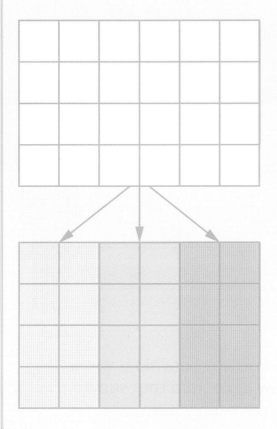

Give a division and a fraction statement for this diagram.

Practise

1 Draw arrays on squared paper to help you find:

   a $\frac{1}{4}$ of 20   b $\frac{1}{5}$ of 20   c $\frac{1}{10}$ of 20

   d $\frac{1}{2}$ of 30   e $\frac{1}{3}$ of 30   f $\frac{1}{6}$ of 30

2 There are 36 children in a Stage 4 class.
  The children named their favourite sports.

  $\frac{1}{4}$ like soccer $\frac{1}{3}$ like tennis $\frac{1}{6}$ like swimming $\frac{1}{9}$ like basketball

  The remaining children like athletics. How many children like each sport?

Try this ★

Look at the diagram and use your **convincing** skills.

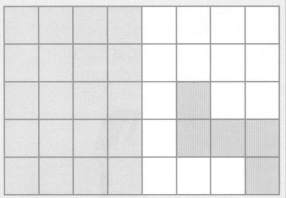

a Which colour shows $\frac{1}{2}$ of 40?

b Which colour shows $\frac{1}{5}$ of 40?

c Which colour shows $\frac{1}{8}$ of 40?

d What fraction does the other colour show?

Let's talk ★

Use your **critiquing** skills.

I have shaded $\frac{1}{4}$ of this shape, because I shaded exactly four squares.

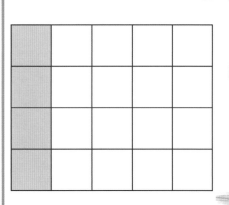

Discuss why Sanchia is wrong. Draw a diagram that shows the correct answer.

More about equivalent fractions

Explore

Look at these shapes.

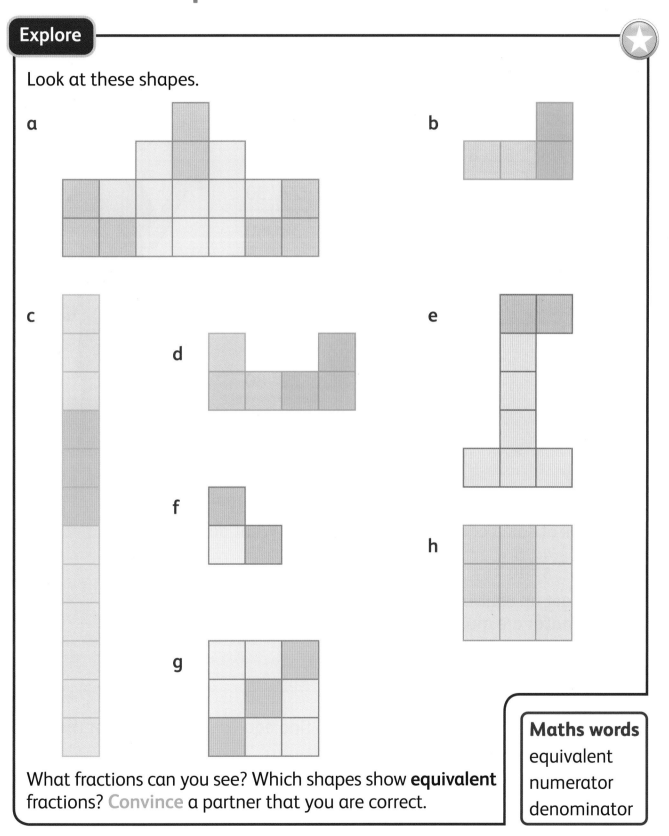

a

b

c

d

e

f

h

g

What fractions can you see? Which shapes show **equivalent** fractions? Convince a partner that you are correct.

Maths words
equivalent
numerator
denominator

Learn

Look at this multiplication grid.

×	1	2	3	4	5	6	7	8	9	10
1	1	2	3	4	5	6	7	8	9	10
2	2	4	6	8	10	12	14	16	18	20
3	3	6	9	12	15	18	21	24	27	30
4	4	8	12	16	20	24	28	32	36	40
5	5	10	15	20	25	30	35	40	45	50
6	6	12	18	24	30	36	42	48	54	60
7	7	14	21	28	35	42	49	56	63	70
8	8	16	24	32	40	48	56	64	72	80
9	9	18	27	36	45	54	63	72	81	90
10	10	20	30	40	50	60	70	80	90	100

Look at the pattern that the multiples of **1** and **2** make.

Now look at some equivalent fractions of $\frac{1}{2}$. What do you notice?

$\frac{1}{2} = \frac{2}{4} = \frac{3}{6} = \frac{4}{8} = \frac{5}{10}$ …

You will get an equivalent fraction when both the **numerator** and **denominator** are multiplied by the same number.

Practise

1 Use the multiplication grid in Learn to find equivalent fractions for these.

$$\frac{1}{4} = \frac{2}{8} = \frac{3}{12} = \bigcirc$$

a $\frac{1}{5}$　　　　b $\frac{1}{10}$　　　　c $\frac{3}{4}$　　　　d $\frac{2}{3}$

Practise *(continued)*

2 Equivalent fractions share the same position on the number line.
   Write the equivalent fractions shown by the shaded pairs.

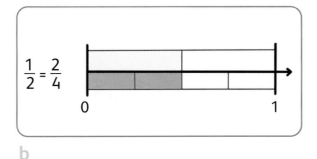

a

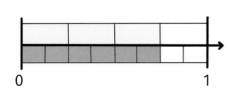

b

c

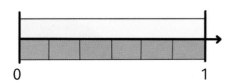

3 A group of five boys share one pack of animal cards.
   A group of 15 girls share some packs of animal cards.
   The girls each have the same fraction of one pack as each of the five boys.
   How many packs did the 15 girls share?
   Draw something to **convince** your partner how you know.

Let's talk

Elok and Pia are playing a game.
They take turns to draw a number
and choose a fraction.

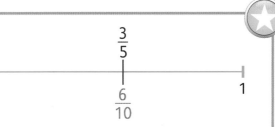

Elok chooses $\frac{3}{5}$ and labels it.

Pia writes an equivalent fraction under the number line.
Can she write more equivalent fractions?
Play Elok and Pia's game with a partner.
How many different equivalent fractions can you use?
Use your skill of **specialising**.

Adding and subtracting fractions

Explore

What different fractions can you see?
Which pairs of fractions make one whole?

Maths words
add
subtract

Learn

We can **add** and **subtract** pairs of fractions.

$\frac{1}{4}$	$\frac{1}{4}$	$\frac{1}{4}$	$\frac{1}{4}$

$\frac{3}{4} + \frac{1}{4} = \frac{4}{4}$ We can write $\frac{4}{4}$ as 1.

The diagram also shows us the related fact: $\frac{4}{4} - \frac{3}{4} = \frac{1}{4}$

Look at the two fractions shown here.

$\frac{1}{5}$	$\frac{1}{5}$	$\frac{1}{5}$	$\frac{1}{5}$	

$\frac{1}{5}$	$\frac{1}{5}$	$\frac{1}{5}$		

What is the same about them? What is different?
Will their total be more than 1? How do you know?

$\frac{4}{5} + \frac{3}{5} = \frac{7}{5}$

$\frac{1}{5}$	$\frac{1}{5}$	$\frac{1}{5}$	$\frac{1}{5}$	$\frac{1}{5}$

$\frac{1}{5}$	$\frac{1}{5}$			

This diagram shows $\frac{4}{5} - \frac{1}{5}$. What is the answer?

$\frac{1}{5}$	$\frac{1}{5}$	$\frac{1}{5}$	𝗫̶	

Practise

1 Draw a diagram each time to show that each pair of fractions totals 1.

   a $\frac{2}{3} + \frac{1}{3}$ b $\frac{5}{8} + \frac{3}{8}$ c $\frac{1}{6} + \frac{5}{6}$ d $\frac{4}{6} + \frac{6}{2}$

2 What additions and subtractions do these diagrams show?
   Write the calculations and answers.

   a    b

   c    d

   e 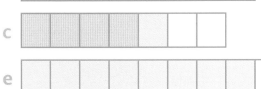   f

Practise *(continued)*

3 Complete these additions. Make an estimate first.
Will the totals be more or less than 1?

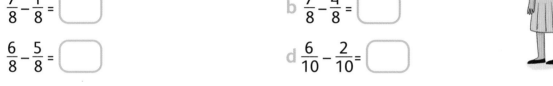

Use your skill of **generalising** while working on questions 3 and 4.

a $\frac{3}{8} + \frac{4}{8} =$ ◯

b $\frac{7}{8} + \frac{2}{8} =$ ◯

c $\frac{1}{4} + \frac{3}{4} + \frac{1}{4} =$ ◯

d $\frac{3}{9} + \frac{2}{9} =$ ◯

4 Complete these subtractions. Will the answers be less than $\frac{1}{2}$?

a $\frac{7}{8} - \frac{1}{8} =$ ◯

b $\frac{7}{8} - \frac{4}{8} =$ ◯

c $\frac{6}{8} - \frac{5}{8} =$ ◯

d $\frac{6}{10} - \frac{2}{10} =$ ◯

5 Solve these problems.

a Pia colours $\frac{3}{8}$ of a shape blue. She then colours another $\frac{2}{8}$ red.
What fraction of the shape is coloured in total?

b Banko has a bag of marbles. He gives $\frac{1}{4}$ of the marbles to Pia.
What fraction of the bag of marbles does he have left?

c Elok wants to buy a new game.
In the first week, she saves $\frac{2}{5}$ of the money she needs.
In the second week she saves $\frac{2}{5}$ more.
What fraction of the total amount of money does she still need to save?

Try this

Choose to add together two or three of these fractions.

$\frac{2}{10}$ $\frac{4}{10}$ $\frac{1}{10}$ $\frac{3}{10}$ $\frac{5}{10}$

Draw diagrams like this to show different totals.

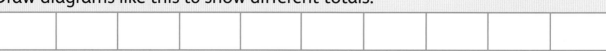

Estimate if the total will be more than 1.

Quiz

1 The baker divides three cakes equally between four plates.
  What fraction of a whole cake is on each plate?

2 Look at this array.

Find the following fractions of the total
number of squares:

a $\frac{1}{2}$ b $\frac{1}{4}$ c $\frac{1}{3}$

d $\frac{1}{6}$ e $\frac{1}{12}$

3 Which two shapes show equivalent fractions? Write the two fractions.

a

b

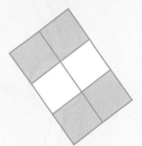

c

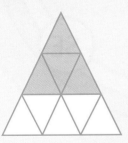

d

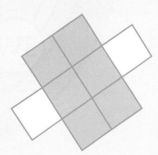

4 Copy and complete.

a $\frac{2}{5} + \frac{3}{5} =$ ⬜   b $\frac{2}{5} + \frac{4}{5} =$ ⬜

c $\frac{4}{5} + \frac{3}{5} =$ ⬜   d $\frac{7}{8} - \frac{2}{8} =$ ⬜

e $\frac{7}{8} - \frac{3}{8} =$ ⬜   f $\frac{7}{8} - \frac{5}{8} =$ ⬜

North, south, east, west

Explore

What is the name of this instrument?

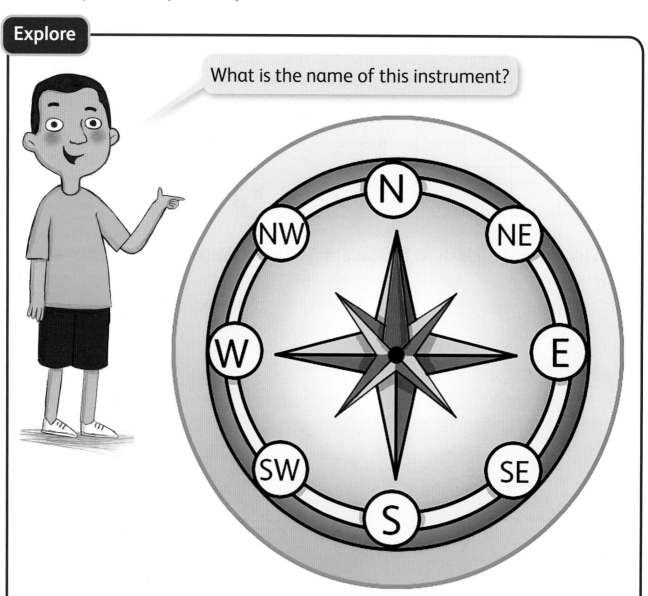

What are the directions NE, SE, SW and NW?

Play some turning games.
Start by facing **north** each time.
Then make half turns and quarter turns to face
east, **south** and **west**.

Maths words

north
east
south
west

Learn

The child is facing north. He turns to face east.

The turn from north to east is a quarter turn.
What angle is this?
How many quarter turns must the child
make to move from east to west?
What is the other name for this turn?

Now imagine that Guss is facing west.
He makes a turn that is less than a right angle.
Discuss the direction that Guss might face now.
Use the picture in **Explore** to help you.

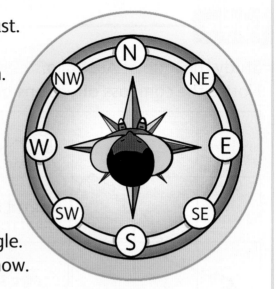

Practise

1 Look at the picture on the right.
  In which direction could the child
  face each time?

  a He makes two quarter turns.
  b He turns by a right angle. Think
    about clockwise or anticlockwise.
  c How many degrees must he turn
    to be facing east again?

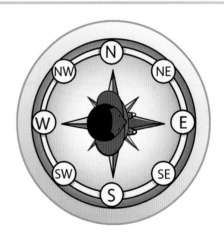

2 a Copy the picture on the right.
    Write the name of the shape
    at each direction.

    • North
    • East
    • South-east
  b In which direction is the
    moon shape?

North

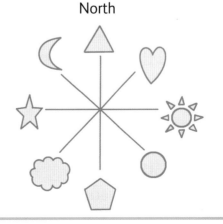

Directions and maps

Explore

Study the map of Australia.
Notice the compass **directions** and the **scale**.

Key	
☐☐	= 1 000 km

Indian Ocean

Darwin

South Pacific Ocean

NORTHERN TERRITORY

Great Barrier Reef

Broome

Cairns

Alice Springs

Uluru

QUEENSLAND

WESTERN AUSTRALIA

SOUTHERN AUSTRALIA

Perth

NEW SOUTH WALES

Sydney

Canberra

Adelaide

Southern Ocean

VICTORIA

Melbourne

TASMANIA

N
NW   NE
W   E
SW   SE
S

What is to the north, east, south and west of Australia?
Describe the position of the different oceans.
Estimate the width of Australia.
Estimate the length of Australia, from the tip of
Queensland to the bottom of Tasmania.
Research other maps.
Use **north**, **east**, **south** and **west** to describe places.

Maths words

direction

scale

Use the **Explore** map to find out what you can see at the nature reserve.

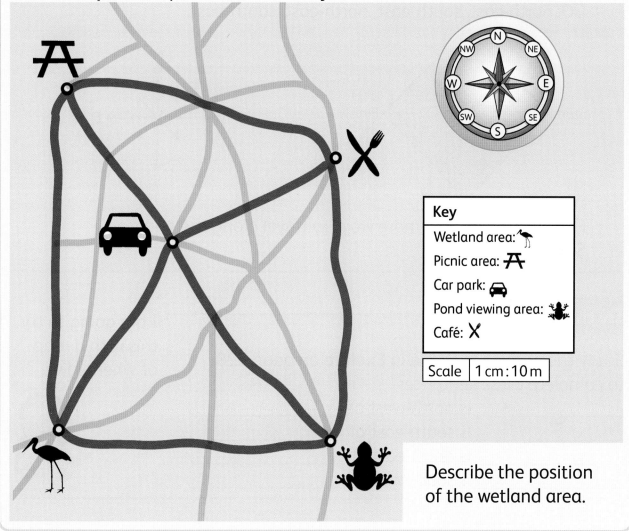

Key

Wetland area:
Picnic area:
Car park:
Pond viewing area:
Café: X

Scale | 1 cm : 10 m

Describe the position of the wetland area.

Practise

 1 Study the map of the nature reserve. Use your critiquing skills.

 a If you stand at the pond-viewing area, what direction must you face to see the wetland area?

 b Describe the direction and distance of the walk from the car park to the picnic area.

 c Give directions to get from the cafe to the picnic area to a friend. Use the woodland walk.

Practise *(continued)*

2 a Follow this description of the journey along the grid from Start to Finish.
GO: north-east, south-east, north-east, south-east.

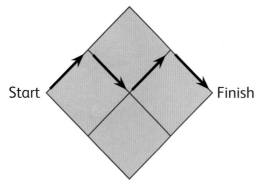

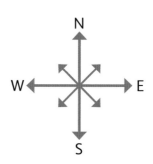

b Find and record four more ways to travel along the grid from Start to Finish.

Try this

I am going to try this with a map of where I live.

Study the map of Australia in **Explore** on page 128 to complete these sentences.

_____ is north-west of _____.
_____ is south-east of _____.
_____ is north-east of _____.
_____ is south-west of _____.

Let's talk

A game for two players
Player A puts their counter on one circle.
Player B puts their counter on another circle.
Player A must try to catch Player B.
Every move must be along one line.
You must say the direction before you move.
Player B moves first.

Coordinates

Explore

Look at this pattern.
Describe how you think you can make it.
Look at the way the straight lines seem to make a curve.
Try to make your own pattern like this.

Maths words
coordinate
x-axis
y-axis

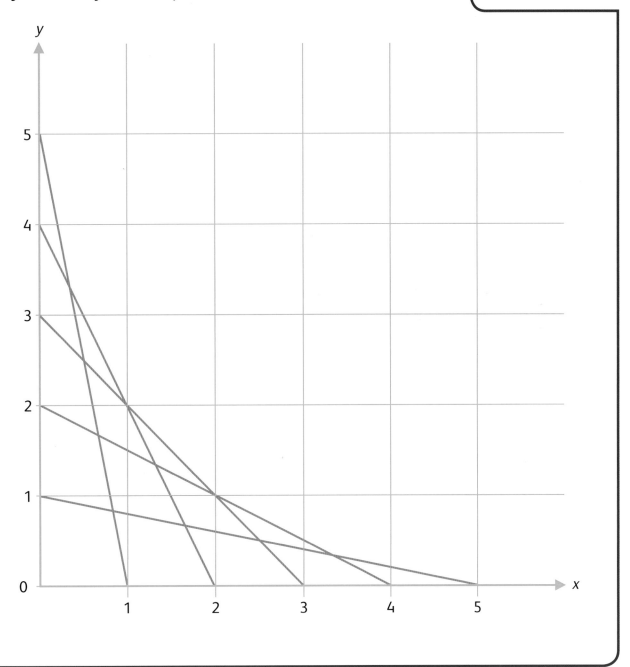

Learn

Look at the diagrams. Try to describe the position of the dot.

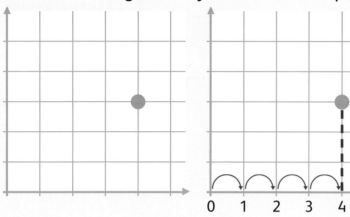 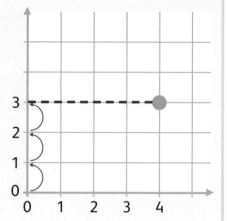

We say:

The dot is **4** along the horizontal axis.

It is **3** up the vertical axis.

The **coordinates** of the dot are (4, 3).

We ALWAYS write the **x-axis** measure first, then the **y-axis** measure.

Use coordinates to describe the position of the star and the triangle.

Compare the coordinates for the star and the circle. What do you notice?

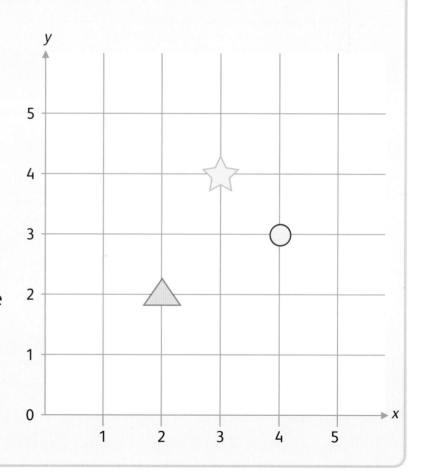

Practise

1 Copy the coordinate grid onto square paper.
Mark dots at these coordinates,
then join them to form a shape.

 (2, 1)   (5, 2)

 (4, 5)   (0, 4)

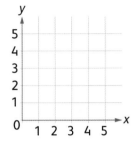

2 Look at the map of some woodland.
Coordinates can help us to find different features.

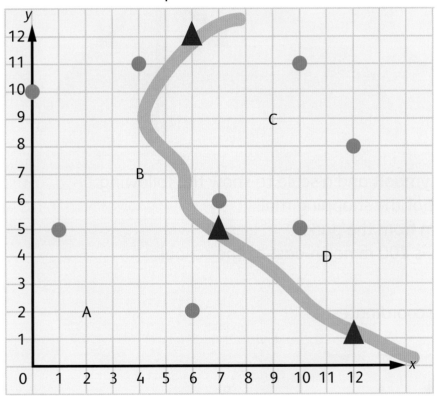

Key

● Tall trees

▲ Bridge over river

a Which of each pair are the coordinates for tall trees?

(2, 6) or (6, 2) (5, 10) or (10, 5) (8, 12) or (12, 8) (0, 10) or (10, 0)

b Write the coordinates for three more trees.

c Guss needs to cross the river. Give coordinates for where he can cross.

d Use your specialising skills. Elok is at B and needs to get to C.
Give directions for travelling along the grid lines.

Quiz

1 Sketch the diagram and add the remaining compass directions.

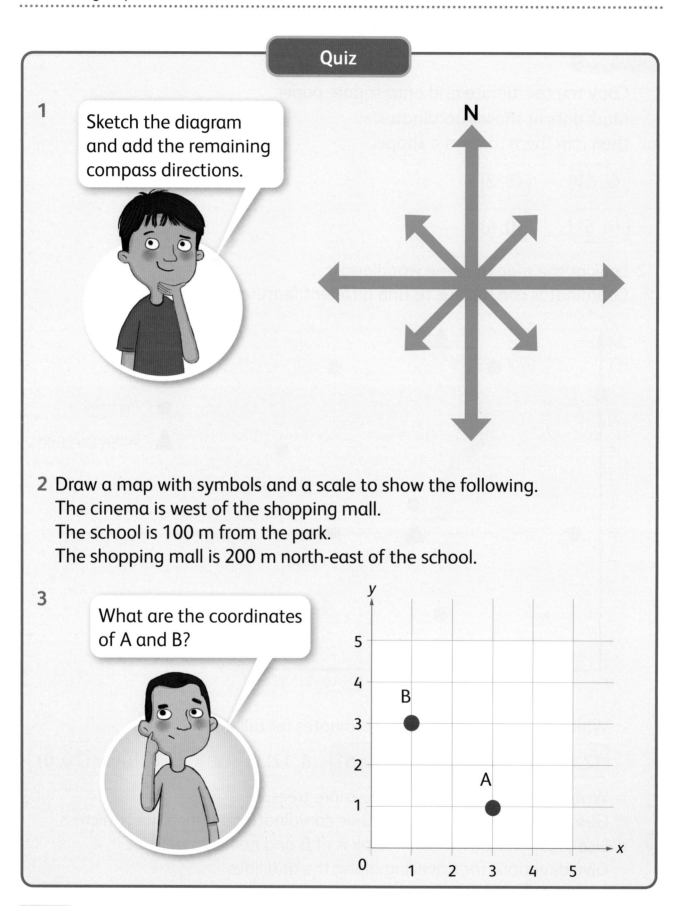

N

2 Draw a map with symbols and a scale to show the following.
   The cinema is west of the shopping mall.
   The school is 100 m from the park.
   The shopping mall is 200 m north-east of the school.

3 What are the coordinates of A and B?

Units 7–12

1 Work out the cost of the bat and the ball.

a [bat icon $?] + [bat icon $?] = $18   b $20 − [ball $?] = $14

2 Copy and complete.

a 30 × 4 = ◯   b 36 × 4 = ◯

c 400 × 7 = ◯   d 423 × 7 = ◯

3 Choose probability words to complete each sentence.

a It is _____ that the next car I see will have four wheels.

b The probability of a thunderstorm in the next five minutes is _____.

4 Find the following fractions of 24.

a $\frac{1}{2}$ = ◯   b $\frac{1}{3}$ = ◯

c $\frac{1}{4}$ = ◯   d $\frac{1}{6}$ = ◯

e $\frac{1}{8}$ = ◯

5 Copy and complete.

a $\frac{3}{10} + \frac{2}{10}$ = ◯   b $\frac{5}{8} +$ ◯ $= \frac{8}{8}$

c $\frac{3}{5} + \frac{4}{5}$ = ◯   d $\frac{9}{10} - \frac{5}{10}$ = ◯

e $\frac{7}{9} -$ ◯ $= \frac{5}{9}$

6 Draw a shape with an acute angle.

7 Draw a shape with two obtuse angles.

8 What is the next term in each sequence?
Write the next term and the term-to-term rule each time.

a 18, 12, 6, 0, _____

b 2, 4, 8, 16, _____

c 88, 44, 22, _____

9 Copy and complete.

a 6 and 4 are factors of 24 because ☐ × ☐ = ☐

b _____ is a multiple of _____ and _____ because 24 ÷ 3 = 8

10

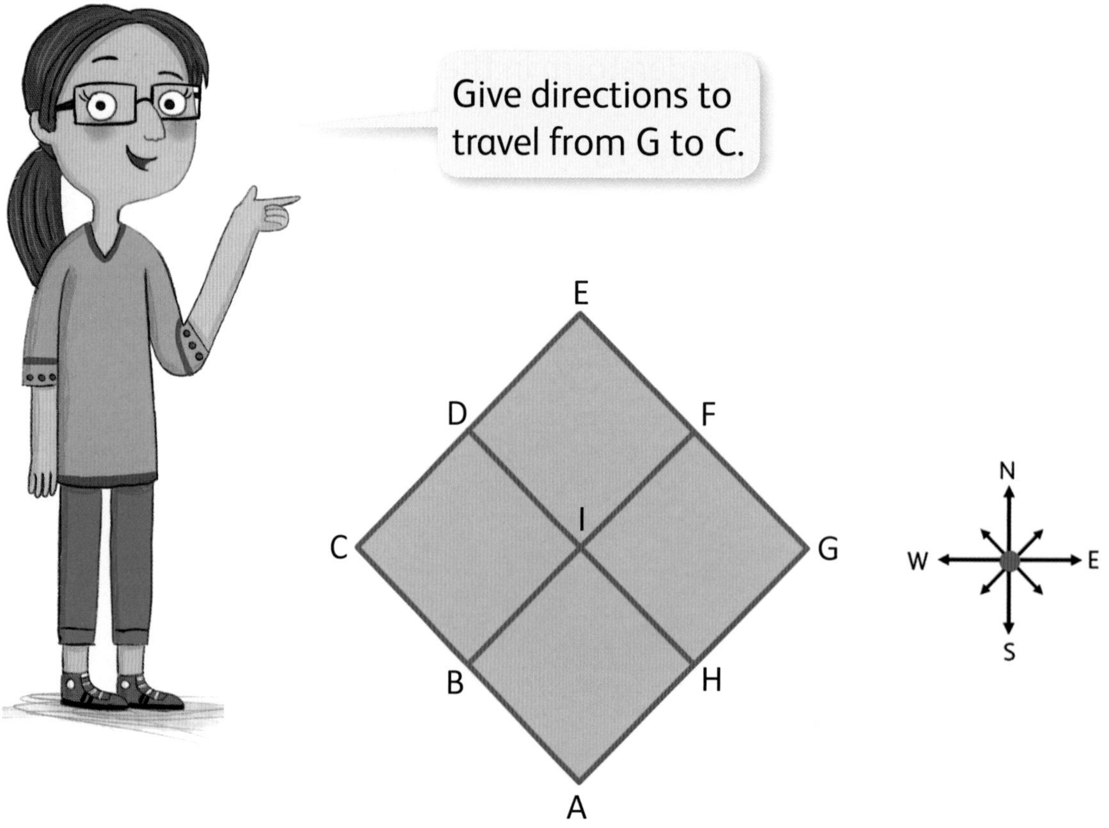
Give directions to travel from G to C.
E
D
F
C
I
G
B
H
A
N
W
E
S

Identifying and building sequences

Explore

The thermometers show a **sequence** of temperatures, both negative and positive.

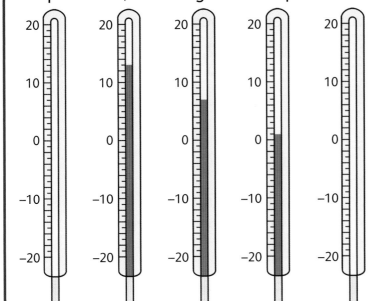

Maths words

sequence   rule
recursion   term

What do you notice about the temperatures?
What will the missing temperatures be?
Explain your thinking.

Learn

Patterns of numbers that follow a **rule** are called sequences.
Let's use the **recursion** rule **multiply by 10 and add 1**,
to build a sequence starting from **2**.

What are the missing numbers in this sequence?

A term-to-term rule is also called recursion.

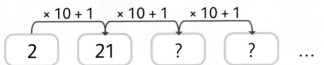

What is the value of the fifth term in the sequence?
Will it be even or odd? Why?

Practise

1 Use the rule to build each sequence. Write the next four terms.

	Rule	First term
a	Add 1 and multiply by 10	2
b	Divide by 2 and add 4	248
c	Multiply by 5 and subtract 3	1

2 Write the rule and the missing values for each sequence.

a 3, 6 , 12, 24, ☐ , ☐ , ☐   b 3, −2, −7, ☐ , ☐ , ☐

c 30, 60, 120, 240, ☐ , ☐ , ☐   d ☐ , 20, 10, 0, ☐ , ☐

e ☐ , 8 000, 4 000, 2 000, ☐ , ☐

f 20 000, 2 000, 200, ☐ , ☐

3 The sequence is represented as a pattern. Use your conjecturing skills. What is the rule? How many dots will be in the next two terms? Sketch the possible patterns.

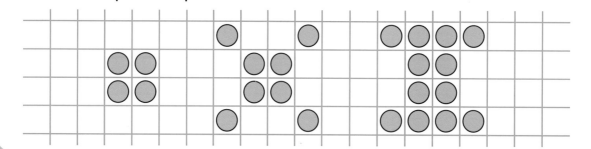

Try this

Pia is making up different sequences. She uses **5** as the start number each time.

The first four **terms** in each sequence are the same when I use the rule **add 4 and subtract 3** or **subtract 3 and add 4**.

Is Pia correct? Critique her work.
What else do you notice? Can you explain why?

Square numbers

Explore

Look at this arrangement of oranges.

Maths words
array
square number
factor pair

What do you notice about each layer?
Each layer is like an **array** with equal rows and columns.

How many oranges in each layer?
How many oranges in total?

Learn

The **arrays** represent the top three layers of oranges.

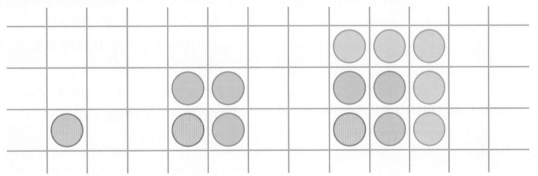

We say that the numbers 1, 4 and 9 are **square numbers**. Can you think why?
Use the arrays to describe how the next square number is built up from the previous square number.
What do you notice about the rows and columns?
A square number is the result of multiplying a number by itself.
One of its **factor pairs** will be made up of the same two numbers.
For example, 3 and 3 are a factor pair of 9.

Practise

1 Use counters to build the next three square numbers after 9.
   Write the matching multiplication facts.

2 a Find all the square numbers up to 10 × 10.
     Use the multiplication facts you know.
   b List the sequence of square numbers you have found, starting from 1.

3

I have eight bags
of seven counters.

I have nine bags
of four counters.

I have five bags
of six counters.

a Which child can make
  a square array using all
  of their counters?

b How many rows and how
  many columns will the
  array have?

Let's talk

Only one of these five numbers is square.

| 18 | 24 | 35 | 49 | 60 |

Use the following to explain how you know.
• Arrays
• Multiplication facts
• Factor pairs

Remember to show why the
other numbers are not square.

Try this

Write all the factor pairs for 4.
Now write all the factor pairs
for 16 and then for 25.
Make a generalisation.

Factors and multiples

Explore

Sanchia and Jin are making a **factor pair** diagram for 24.

Critique what they have done so far. Do you agree?

What pairs of **factors** are missing?

Explain why 24 is not a square number.

Maths words
factor pair
factor
multiple

Learn

24 is a **multiple** of 3 and 8 because 24 ÷ 3 = 8 and 24 ÷ 8 = 3.

3 and 8 are factors of 24 because 3 × 8 = 24.

Describe other factor pairs of 24 in a similar way.
We know that 5 is not a factor of 24 because 5 does not divide exactly into 24.
It will leave a remainder. 24 is not in the 5× multiplication table.

Practise

1 Write all the factor pairs for each number.

a 15   b 21   c 28   d 36

2 Look at the patterns of multiples in the clouds.

> Multiples of 3
> 0, 3, 6, 9, ___

> Multiples of 4
> 0, 4, 8, 12, ___

a Continue each pattern of multiples, but do not go past the number 50.

b What is the largest number you have that is a multiple of both 3 and 4?

c Are there any numbers in each list that are also multiples of 5?

 3 Use your classifying skills.

a Identify the incorrect factor each time. How do you know?

Factors of 25: 5, 1, 3   Factors of 28: 9, 2, 7

Factors of 40: 10, 8, 6   Factors of 100: 21, 100, 4

b Add another factor for each of 25, 28, 40 and 100.

Try this

I know that 2 is a factor of 10, 16 and 24.

I know that 15, 30 and 45 are multiples of 5.

Are Elok and Banko correct? How do you know? What other numbers could they have in their lists? Explain your ideas.

Let's talk

True or false? Convince your partner of your thinking.

25 is a multiple of 2.

4 and 8 are a factor pair of 32.

3 and 3 are a factor pair of 10.

9 is a factor of 27.

Make up some other true or false statements for others to solve.

Tests of divisibility

Explore

Here are some patterns showing **multiples** of three different numbers.

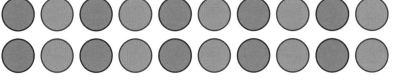

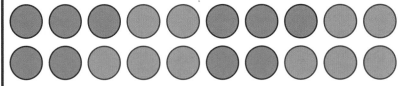

What numbers are they?
What multiples can you see?
What else do you notice?

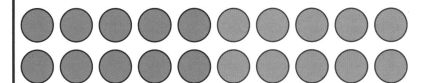

Maths words
multiple
divisible
factor

Learn

All multiples of **two** are **even** numbers.
They have the ones digit 0, 2, 4, 6 or 8, for example, 3**6** and 4**8**.
Multiples of **five** have the ones digit **0 or 5**, for example, 4**5** and 5**0**.
Multiples of **ten** have the ones digit **0**, for example, 6**0** and 10**0**.
Multiples of ten are also multiples of 2 and 5. Try to think why.

We can use these tests of divisibility to identify any number that is **divisible** by 2, 5 or 10.

Practise

1 a Use the tests of divisibility to decide which numbers are divisible by
2, 5 or 10: 76   95   120   324   429   645   700

Write them in a table like this.

Divisible by 2	Divisible by 5	Divisible by 10

b One number is left over. Explain how you know that this number is
not divisible by 2, 5 or 10.

2 For each question below, write five numbers that you can place in the
shaded part of this number line.

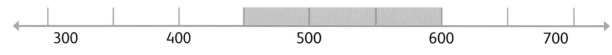

a Numbers that are divisible by 2   b Numbers that are divisible by 5
c Numbers that are divisible by 10

3 Decide if these statements are **always true**, **sometimes true** or **never true**.
a Multiples of 2 are divisible by 2.
b Odd numbers are multiples of 2.
c Numbers that are divisible by 5 are multiples of 2.
d Multiples of 10 are divisible by 10.

Let's talk

All multiples of 10 are also multiples of 5, so
all multiples of 5 must also be multiples of 10.

Critique Sanchia's thinking.
Use tests of divisibility to help you explain the mistake Sanchia has made.

Learn

We can use what we know about **factors** and multiples to help us understand other tests of divisibility.

Where can you see multiples of 25?
Where can you see multiples of 50?
How does the picture help you to think of factors of 100?

What is the connection between 5 and 10, and 50 and 100?

Practise

1 a Write the multiples of 25 up to 250.
   b Write the multiples of 50 up to 500.
   c Write the multiples of 100 up to 1 000.
   d What patterns do you notice? Write some of your ideas.

2 Look at these numbers and classify them.

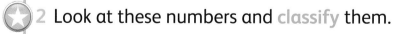

Which numbers are divisible by these numbers?
   a Divisible by 100
   b Divisible by 50
   c Divisible by 25

Practise

3 A farmer has 400 apple trees and 350 pear trees to plant.
  a Can he plant the apple trees in rows of 25, 50 or 100?
    How do you know?
  b Can he plant the pear trees in rows of 25, 50 or 100?
    How do you know?

Try this

The amount of money in the car money box is divisible by 25.
The amount of money in the house money box is a multiple of 50 cents.
The amount of money in the football money box is divisible by 100.

Use your skill of specialising. Find three solutions so that:
● The amount of money in each box is different.
● The amount of money in two boxes is the same,
  but the amount of money in the third box is different.
● All three amounts of money are the same.

Remember, in Learn on page 145, you found out about the connection between different numbers.

Quiz

1 What is the term-to-term rule for these sequences?
Write the values of the missing terms.

a 75, 50, 25, 0, ⬜

b ⬜ , 1, 2, 4, ⬜ , ⬜

2 A sequence follows the rule: Multiply by 5 and subtract 5.
The first term is 2. Write the next three terms in the sequence.

3 Which of these numbers will make a square array?

| 14 | 24 | 34 | 44 | 54 | 64 |

4 Why will the answer to 5 × 5 be a square number?

5 Find the missing numbers to complete these factor pairs.

a 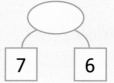   b 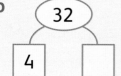   c 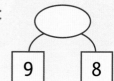   d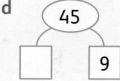

6 Use tests of divisibility to help you explain which is the odd one out
each time.

a | 132 | 274 | 383 | 456 |

b | 205 | 230 | 255 | 262 |

c | 340 | 365 | 380 | 400 |

7 Write two numbers that are greater than 100 for each of these.
  a Divisible by 25
  b Divisible by 25 and 50
  c Divisible by 25, 50 and 100

Interpreting and comparing data

Explore ★

Which letter is most common? Sanchia collects this **data** from a list of first names of learners in her class. Pia collects this data from a page in her reading book.

Investigate the question in your own class. Make **conjectures** about whether you think the results will be similar.

What is the **frequency** of the most common letter? How often does it appear? Decide how to collect the data, **interpret** or understand it and present it to the class.

Learn ★

Compare and contrast these two bar **charts**.

Chart A Class 4

Frequency

14
12
10
8
6
4
2
0

1   2   3   4   5   6   7   8   9   10 or more

Number of letters in learners' first names

Chart B Whole school

Frequency

80
70
60
50
40
30
20
10
0

1   2   3   4   5   6   7   8   9   10 or more

Number of letters in learners' first names

Maths words

data	frequency
interpret	compare
chart	investigate
scale	

Practise

1 Look at Chart A in Learn and answer these questions.
   How many names have six letters?

   > (2) names have six letters.

   a What is the most common number of letters in a name?
   b How many names have more than five letters?
   c How many names have fewer than four letters?
   d How many learners are in Class 4?

2 Look at Chart B in Learn and answer these questions.
   a How many names have five letters?
   b How many names have more than six letters?
   c How many names have fewer than five letters?
   d How many learners are in the whole school?
   e How many more names have four letters than six letters?

3 Look at Chart A and Chart B. Write true or false for each statement below.
   Remember: Class 4 is part of the whole school.
   a There are more than 200 learners who are not in Class 4.
   b In Class 4, five-letter names are more common than three-letter names,
      but in the whole school, it is the other way around.
   c All learners who have two letters in their name are in Class 4.
   d More learners in Class 4 have names with ten letters or more, than in
      the rest of the school.

Try this

Investigate (find out about) name lengths in your class.
Decide how to collect, interpret and present the information.
Which kind of chart will you use?
What **scale** will you use, for example, to show the number of letters?

Investigating and collecting data

Explore

Three towns **collect** data about households that recycle glass, plastic and paper. Compare the results.

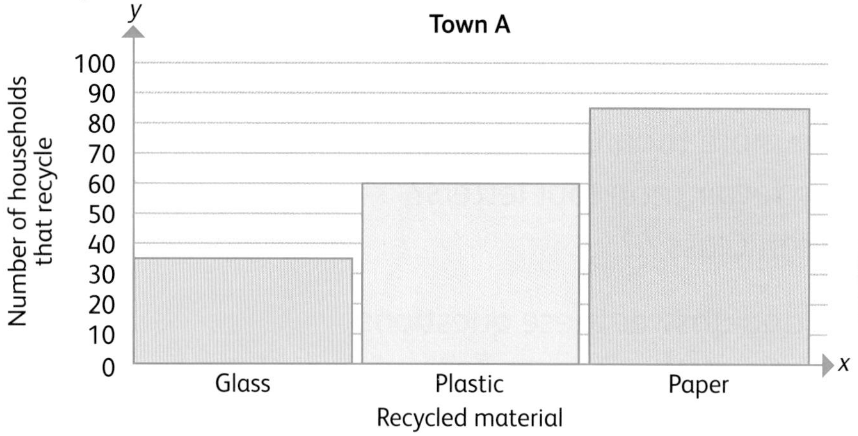

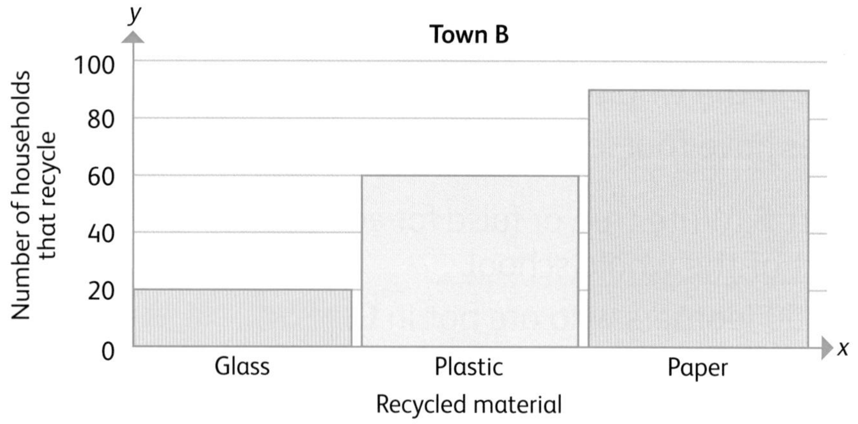

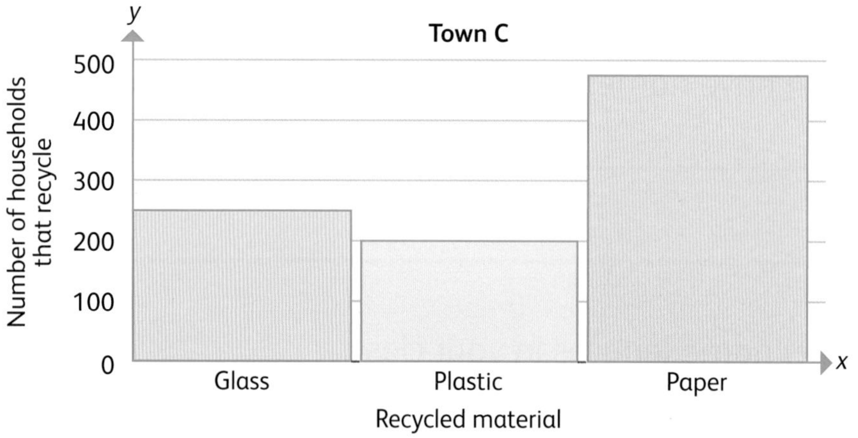

Town A

y — Number of households that recycle

Town B

y — Number of households that recycle

Town C

y — Number of households that recycle

Which town recycles the most paper?

Discuss why there might be differences between the results for the three towns.

Maths words
collect
investigate
decide
interpret

Learn

Collecting data is an important way to **investigate** and solve problems. It is the way scientists and governments **decide** on or make decisions. This flow chart shows the process of using data to investigate something and then **interpret** the data. What skills do you need at each stage?

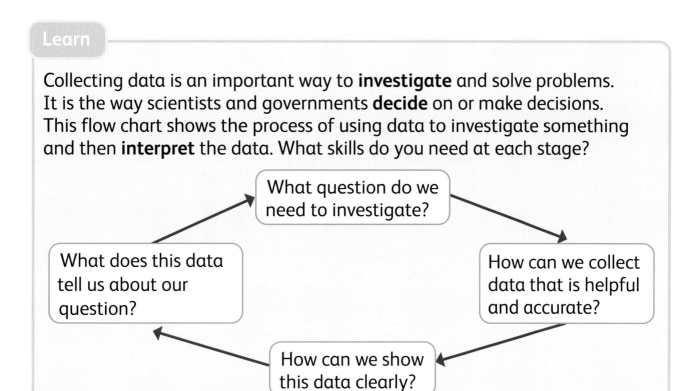

Practise

1 Choose a theme you need or want to investigate, and decide on a question to answer. Here are some examples of themes and questions.

- Recycling. What materials does the school recycle most?
- Saving money. Which resources are wasted at school?
- Improving health. What makes a healthy lifestyle?
  How healthy is your diet? How much sugar do we eat?

2 Decide what information you need to collect.
Design your tables or charts for collecting the information.

Colour	Tally
Red	III
Yellow	HHH IIII
Green	HHH I
Blue	HHH HHH
Orange	II

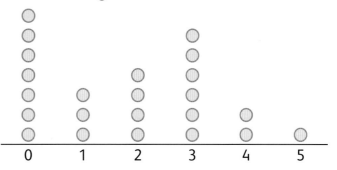

Practise *(continued)*

3 Collect your data. Then choose how to present it to the class.
Make sure that your information is easy to understand.

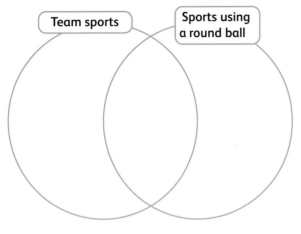

Team sports   Sports using a round ball

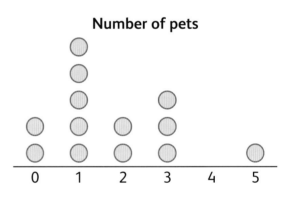

Class 4

Number of letters in learners' first names

Class 2				
Walk	☐	☐		
Car	☐	☐	☐	☐
Bus	☐	▯		
Bicycle	☐			
Key	☐ = 👥			

Number of pets

0   1   2   3   4   5

Try this

Discuss the findings from your investigation.

What are the next steps? What action should you or your school take?

Quiz

1 Explain why you would choose a different scale for different bar charts.

2 Explain how to make sure that you collect accurate data.

More missing number problems

Explore

Elok and Guss have been baking cookies.

I put the same number of cookies in my two tins!

Elok, you baked 12 cookies in total! I baked four cookies more than you.

They put them in tins and write how many cookies there are. Some of the numbers have rubbed off. What are they?

Maths words

symbol   double

half   total

difference

Learn

Symbols have been used to represent the missing numbers in these problems.

$\triangle + \triangle = 20$ $\triangle - 7 = \pentagon$

Can you use the words **double**, **half**, **total** and **difference** to help you describe what you know about the missing numbers?
Find the missing numbers.

Practise

1 Find the value of each symbol.

a ☆ + ☆ = 8

b 20 − ☺ = 15

c − 9 = 11

d 10 − ⬭ = ⬭

2 Find the missing prices.

a   +   = $10

$20   −   =

b   −   = $4

   + $3   = $5

3 Find the missing numbers.

a ☆ + ◯ = 40   50 − ◯ = 20

b ▱ + ▱ = 100   70 − △ = ▱

c ◇ − ⬡ = 13   ⬡ + ⬡ = 14

Try this

Use addition and subtraction facts to 10
to make up some missing number problems.
Think about the symbols you will use.
Convince a partner that they are correct.

Ask a friend
to solve the
problems you
make up!

More addition and subtraction

Explore

The children are playing an **addition** and **subtraction** game.
They pick five digits and put one digit in each box.

They make the addition: **652 + 238**

What is a sensible **estimate**?
Find the total.
The children then subtract the smaller number from the larger number.
Is the answer more or less than 400? Is it even or odd?

Make other calculations using your own digit cards.

Maths words
addition
subtraction
estimate

Learn

Pia and Jin make up some calculations.

| 658 − 197 | 475 + 238 | 614 − 583 | 563 + 279 | 299 + 973 |

They can't decide if they should use mental or written methods to solve them.
Help them to decide. Explain your thinking.
What would be a sensible estimate for each calculation?
Explain whether the actual answers will be more or less than your estimates.

Practise

1 A road sign reads:

Anytown	303 km
Betterville	427 km
Coolton	532 km
Downton	601 km
Everhampton	840 km

How far away will each place be after
the driver drives 285 km more?

2 Pick a number from the square and a number from the triangle.

308   475
283   146

496
658   523   295

What happens when
you add 400 + 600?

Make an addition and a subtraction.
Estimate the answers and then complete
the calculations using a method of your choice.

3 184 more people went to a cricket match than to a football match.
297 fewer people went to a basketball game than a tennis match.
There were 935 people at the football match.
This was 99 more people than were at the tennis match.
How many people were at each match?

Try this

Can you see how these puzzles work?
Find the missing numbers.

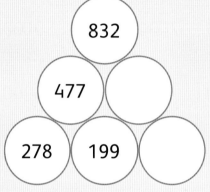

832

477

278   199

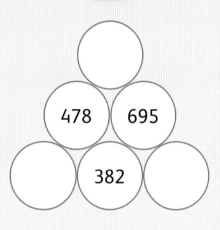

478   695

382

Let's talk

Look at the three subtractions.
Explain any patterns you notice.
Do not work out the answers yet.
Predict which subtractions will have an answer that:

a   ends in a zero

b   is less than 100

c   is between 200 and 300

d   is an even number.

321 – 123

432 – 234

543 – 345

Now do each
calculation to check.

157

Multiplying and dividing whole numbers by 10 and 100

Explore

This is a Gattegno chart.

10 000	20 000	30 000	40 000	50 000	60 000	70 000	80 000	90 000
1 000	2 000	3 000	4 000	5 000	6 000	7 000	⬤	9 000
100	200	300	⬤	500	600	700	800	900
10	20	30	40	50	⬤	70	80	90
1	2	3	4	5	6	7	8	9

What do you notice about the way the numbers are arranged?
How many times as large are the numbers in the middle row
than the numbers in the bottom row?
What numbers are hidden by the counters?
Explain how you know.

What is the total value
of the hidden numbers?

Maths words
place value
multiply
divide

Learn

These **place value** charts show how digits change when we **multiply** a number by 10 or 100. Write a calculation for each diagram.

100s	10s	1s
	3	5
3	5	0

1 000s	100s	10s	1s
		3	5
3	5	0	0

Why are the zeros important? What role does zero have?

Look at the way multiplying by 10 makes a number 10 times as large.

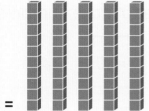 × 10 =

What happens to the size of the number when we multiply it by 100?

Division is the inverse of multiplication.
Can you explain what happens when we **divide** 350 by 10?
What happens when we divide 3 500 by 100?

Practise

1 Copy and find the missing numbers.

a 33 × 10 = ⬚

100s	10s	1s
	3	3

b 550 ÷ 10 = ⬚

100s	10s	1s
5	5	0

c 42 × 100 = ⬚

1 000s	100s	10s	1s
		4	2

d 2 100 ÷ 100 = ⬚

1 000s	100s	10s	1s
2	1	0	0

Practise *(continued)*

2 Copy and complete.

a 13 × 10 = [130]

23 × 10 = [ ]

230 × 10 = [ ]

235 × 10 = [ ]

b [ ] × 10 = 190

[ ] × 10 = 380

[ ] × 10 = 3 800

[ ] × 10 = 3 840

c 450 ÷ 10 = [ ]

540 ÷ 10 = [ ]

[ ] ÷ 10 = 36

[ ] ÷ 10 = 63

3 Use the cards to make up different multiplications and divisions. You may use the digit cards more than once to make numbers such as 55 or 200.

a
| 2 | 5 | 0 | ×10 | ÷10 | = |

b
| 4 | 6 | 0 | ×100 | ÷100 | = |

4 The capacity of a jug is 560 ml.
The capacity of an egg cup is 10 times as small as the jug.
The capacity of a bucket is 100 times as large as the egg cup.

a What is the capacity of the egg cup?

b What is the capacity of the bucket?

Try this

Practise your *specialising* skills.

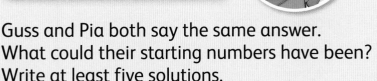

I am thinking of a number.
I multiply it by 10.

I am thinking of a number.
I divide it by 100.

Guss and Pia both say the same answer.
What could their starting numbers have been?
Write at least five solutions.

Simplifying multiplications

What do you notice about these shapes?

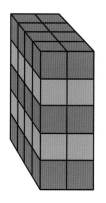
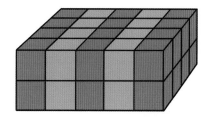
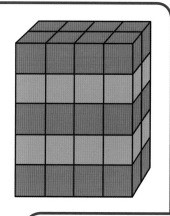
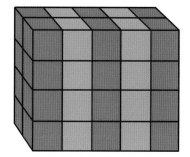

How many small cubes make up each shape?
How can you work it out?

Maths words
multiplication
commutative
associative

Learn

What is the same and what is different about each **multiplication**?

$4 \times 2 \times 5$ $2 \times 4 \times 5$ $5 \times 4 \times 2$ $2 \times 5 \times 4$

We can say: First we multiply 4 and 2. Then we multiply the answer by 5.
Which order was easier? Why?

We can also look for pairs of factors that are easy to multiply.

$4 \times 2 \times 5 = 4 \times 2 \times 5$ 2×5 is a fact we know

$4 \times 10 = 40$ 4×10 is another fact we know

It is useful to look for pairs of factors to help you multiply larger numbers.
We can break up larger numbers into factors that we know.

$8 \times 18 = 8 \times 9 \times 2$ 8×9 is a fact we know

$72 \times 2 = 144$ 72×2 is the same as doubling 72.

Where can you still see 18? Can we also use factors to break down 8×18 as $18 \times 2 \times 2 \times 2$? Why?

Practise

1 Write three different multiplication sentences to match each shape.
  Find the total number of cubes.

a

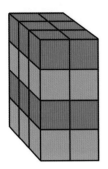

b

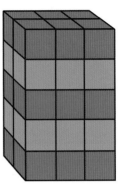

c

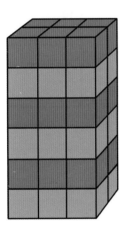

Remember that multiplication is **commutative**, so we can calculate in any order and the result stays the same.

We say that multiplication is **associative** because we can group the factors in different ways and the result stays the same.

2 How can you make these multiplications easier?
  a 2 × 9 × 5
  b 2 × 6 × 3
  c 6 × 5 × 2
  d 5 × 3 × 4

Practise *(continued)*

3 Use factor pairs to make these calculations easier.

$$40 \times 9 = 4 \times 10 \times 9 = 4 \times 9 \times 10$$
$$36 \times 10 = \mathbf{360}$$

a 28 × 5

b 8 × 16

c 45 × 4

d 32 × 5

e 7 × 18

Try this

Jin uses the factors 9, 4 and 5 to help him solve a multiplication:

10s	1s		1s
		×	

What could the multiplication be? Use your skill of specialising to find two possible solutions.

Let's talk

Use cubes to make shapes that match these multiplication sentences.

4 × 5 × 3   6 × 2 × 4

3 × 7 × 2

What other multiplication sentences does each shape represent? Convince you partner why the answers are the same each time.

Multiplying larger numbers

Explore

The children are working with kitchen scales.

Maths word
product

My shapes weigh more in total because each one has a greater mass.

Cuboid 156 g
Pyramid 117 g
Cylinder 134 g

My shapes weigh more in total because there are more of them.

Use your skill of critiquing to say who is correct.
Use estimates to help you.

Learn

The mass of a cone is 237 g.
What will the mass of four of these cones be?

A useful estimate is 200 × 4.
How can we work out the actual mass?
Will the actual mass be more or less than the estimate?
We can solve the multiplication using a mental or written method.

Mental method	Written method

Mental method

	200	30	7
4	800	120	28

800 + 120 + 28 = **948**

Written method

	100s	10s	1s	Method
	2	3	7	7 ones × 4 = 28 ones or 2 tens and 8 ones
×			4	3 tens × 4 = 12 tens plus 2 tens is 14 tens or 1 hundreds and 4 tens
	9	**4**	**8**	2 hundreds × 4 = 8 hundreds plus 1 more hundred is 9 hundreds
	1	2		

Multiplication is commutative, so we can calculate in any order and
the **product** remains the same: 237 × 4 = 4 × 237

The product is the result of
multiplying factors together.
For example:
948 is the product of 237 × 4.
We can check the answer using
our estimate or a calculator.

Practise

1 Answer these. Choose to use mental or written methods.
  Make an estimate first.

  a $64 × 6 =$ ☐   b $6 × 164 =$ ☐

  c $97 × 8 =$ ☐   d $8 × 117 =$ ☐

  e $123 × 4 =$ ☐   f $123 × 8 =$ ☐

 2 **Critique** and then **convince**. Use estimates to check these multiplications.

$128 × 3 = 384$	$128 × 6 = 668$	$139 × 5 = 699$
$7 × 93 = 727$	$67 × 8 = 536$	$9 × 98 = 1002$

  a List any that you know are wrong from your estimates
    and explain why.

  b Now use a method of your choice to check that the answers
    to the other calculations are correct.

3 Solve. Make an estimate first.

  a One bag holds 124 marbles.
    How many marbles in eight bags?

  b A large box has a mass of 45 kg.
    What is the total mass of five
    of these boxes?

  c Pia runs seven laps of 142 metres.
    Sanchia runs six laps of 164 metres.
    Who ran further, Pia or Sanchia?

> You can check your answers using
> your estimates or a calculator.

Try this

What are the missing digits?

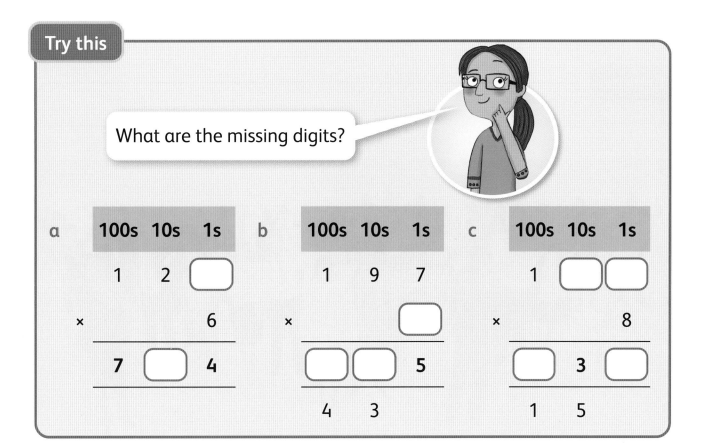

a
100s	10s	1s
1	2	☐
×		6
7	☐	4

b
100s	10s	1s
1	9	7
×		☐
☐	☐	5
	4	3

c
100s	10s	1s
1	☐	☐
×		8
☐	3	☐
	1	5

Let's talk

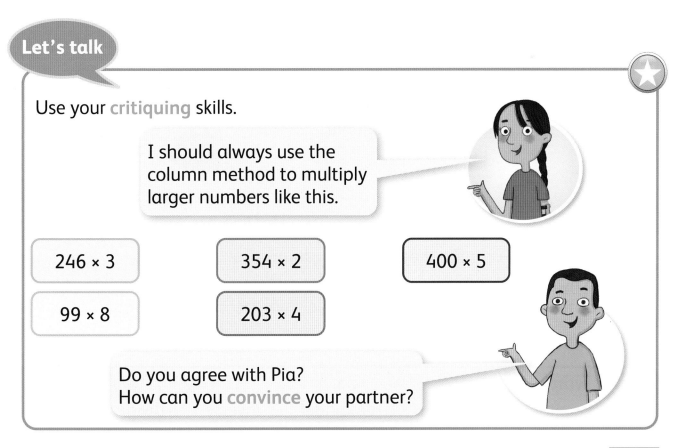

Use your critiquing skills.

I should always use the column method to multiply larger numbers like this.

246 × 3

354 × 2

400 × 5

99 × 8

203 × 4

Do you agree with Pia?
How can you convince your partner?

Dividing 2-digit numbers

Explore

There are 49 people in this picture.

How can they arrange themselves in **equal** groups?
Can all 49 people arrange themselves in **groups** of 2, 5 or 10 with no one left out?
What other **divisions** can you make where some people are left out?

Maths words

equal	group	division	remainder	inverse	divide

Learn

What is a sensible estimate for 56 ÷ 4?

Why do we know that the answer will be more than 10?

The first array shows 56 ÷ 4.

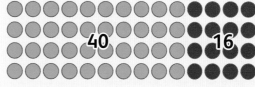

What does the second array show?

What is different about it?

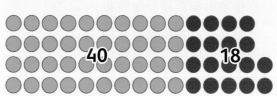

What do you notice about the way each array has been regrouped?

Mental method	Written method
$40 \div 4 = 10$ $56 \div 4$ $16 \div 4 = \dfrac{4}{14} +$	$$\begin{array}{c} \quad\; 1 \qquad 4 \\ \hline 4 \;\big|\; 5 \quad {}^{1}6 \end{array}$$ 1 group of 4 tens → We can now make 4 groups of 4 ones.
$40 \div 4 = 10$ $58 \div 4$ $18 \div 4 = 4 \text{ r } 2$ $58 \div 4 = \mathbf{14 \text{ r } 2}$	$$\begin{array}{c} \quad\; 1 \qquad 4 \quad \text{r } 2 \\ \hline 4 \;\big|\; 5 \quad {}^{1}8 \end{array}$$ 1 group of 4 tens → We can now make 4 groups of 4 ones. There is a **remainder** of 2.

Practise

1 Write and solve the division represented by each array.

a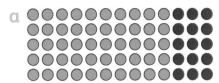

c

b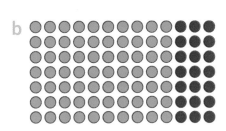

2 Divide 75 by each number. Make estimates first.
   Explain which divisions will definitely leave a remainder.

   a 2   b 3   c 4   d 5   e 6   f 8

3 Banko sketches arrays to show divisions. Write the divisions he is solving.

   a   b   c
       10   6   10   2   30   3
   6 | | | 8 | | | 3 | | |

4 There are 92 tins in a box. The tins are divided equally to go on four shelves.
   How many tins on each shelf?

Try this

64 ÷ 4 = 14
But 14 × 4 = 56

I know my division is wrong because I used multiplication to check it.

51 ÷ 3 = 18   52 ÷ 4 = 13
60 ÷ 4 = 16   96 ÷ 6 = 16
128 ÷ 8 = 16

Use the inverse to check if each of the other divisions is correct.
Now check with a calculator.

Let's talk

Pia thinks of a number between 50 and 100.
When she divides it by 7, the remainder is 2.
It divides by 3 exactly.
What is the number?
Find all the possibilities.

Quiz

1 Find the value of each symbol.

a ▢ + ▢ = 60

b △ − 9 = 20

2 What are the values of the two symbols?

⌓ + ⬠ = 18

20 − ⬠ = 10

3 Copy and complete.

a 654 + 386 = ▢

b 245 + 989 = ▢

c 532 − 142 = ▢

d 843 − 274 = ▢

4 True or false? Correct any that are false.

a 36 × 10 = 360

b 3 600 ÷ 100 = 360

c 63 × 100 = 63 000

d 630 ÷ 10 = 63

5 Use factor pairs to make these calculations easier.

a 14 × 5

b 32 × 5

c 48 × 4

6 a Sort these multiplications from smallest to largest.
Will the products be more or less than 800?

236 × 4

84 × 9

135 × 5

137 × 6

b Now complete the multiplications in part **a**.

7 Copy and complete.

a 68 ÷ 4 = ▢

b 57 ÷ 7 = ▢

c 26 ÷ 3 = ▢

d 48 ÷ 9 = ▢

e 96 ÷ 6 = ▢

Duration

Explore

> **Maths words**
> duration   timetable
> 12-hour   24-hour

The table shows the results of a running race.

Runner A	Runner B	Runner C	Runner D	Runner E
25 s	49 s	36 s	59 s	60 s

Which runner took exactly 1 minute?
Order the runners from 1st to 5th.
What was the difference in time between 1st place and 2nd place?
What about the time difference between 2nd place and 4th place?

Learn

The length of time something takes from start to finish is called its **duration**.
A class **timetable** shows the times when an art lesson starts and ends.

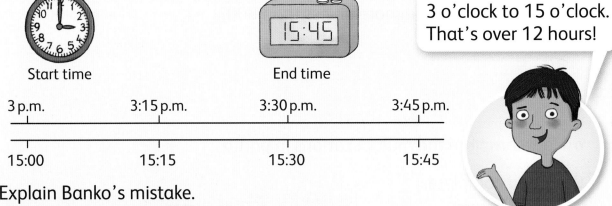

Start time   End time

> 3 o'clock to 15 o'clock.
> That's over 12 hours!

3 p.m.   3:15 p.m.   3:30 p.m.   3:45 p.m.

15:00   15:15   15:30   15:45

Explain Banko's mistake.
How long do you think the art lesson lasted?
Think about the **12-hour** and **24-hour** clocks.

A swimming lesson begins at ten past four after school.
It lasts for half an hour. What time does it finish? Is the number line correct?

16:00   16:10   16:20   16:30   16:40   16:50   17:00

Practise

1 Write the missing times for this TV schedule.

Show	Start	End	Duration
News	09:15	09:45	
Cookery	13:20	13:55	
Sport	15:10		40 minutes
Comedy	17:20		Half an hour
Cartoon		18:30	25 minutes

2 An athlete goes for a run in the afternoon.
He looks at the clock before he leaves.
This map shows his route.

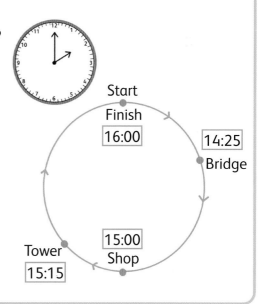

  a What time did the athlete begin running?
  b What time did his run finish?
  c How long did it take him in total?
  d How long did it take him to run each
    of these distances?
    • From the start to the bridge
    • From the bridge to the shop
    • From the shop to the tower
    • From the tower to the end of the run

Try this

There are 60 seconds in one minute.
Calculate these conversions.

  2 minutes = ⬚ seconds

  5 minutes = ⬚ seconds

  10 minutes = ⬚ seconds

  360 seconds = ⬚ minutes

  90 seconds = ⬚ minutes

  100 seconds = ⬚ minutes

Let's talk

A doctor looks at her clock while
working at the hospital.
What time does it show?
What will the time be in 1 minute?

Days, weeks, months and years

Explore

Look at the **calendar**. It shows the **month** of June.

Maths words
calendar
month

JUNE						
Sun	Mon	Tues	Wed	Thurs	Fri	Sat
				1	2	3 ☆
4	5 ▲	6 ☆	7	8	9 ●	10
11	12	13	14	15 ☆	16	17
18	19	20	21	22	23	24
25	26	27 ☆	28	29	30	

Key

☆	Cricket match
●	Pia's party
▲	Jack's party
☐	School holiday

How many days are between Jack's party and Pia's party?
What is the date 7 days after Jack's party?
What is the date 2 weeks after Pia's party?
How long is the longest gap between cricket matches?
The cricket final is 1 week after the last match in June.
What is the date of the final?

Try this

Work out how many days, weeks and months between your birthday and some friends' birthdays.

Try writing the length of time in different units.

Learn

Some mountain explorers recorded their journey in this timetable.
The duration of the journey from the start to Basecamp 1 is longer than a day.

	Leave	Arrive
Journey to Basecamp 1	1 January, 7:00 p.m.	3 January, 7:30 p.m.
Journey to Basecamp 2	4 January, midday	7 January, 3 p.m.
Journey to destination	8 January, 8:20 a.m.	9 January, 8:20 p.m.

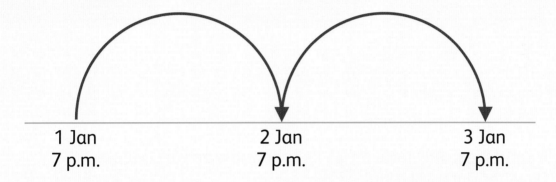

The explorers left for their journey at 7:00 p.m. on 1 January.
They arrived at Basecamp 1 at 7:30 p.m. on 3 January.
That was another 30 minutes.

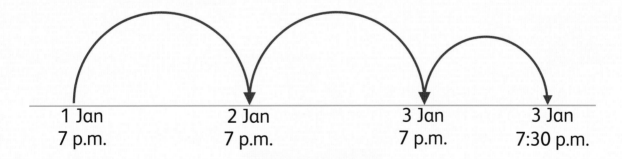

The first stage of the journey was from the start to Basecamp 1.
It took the explorers two days and 30 minutes.
How long did each remaining stage take?
How long did their whole journey take?

Practise

1 This table shows the stages of a building project from start to end.
   Calculate the duration of each stage.

Project	Start	End
Dig ground	10 May 09:00	11 May 11:00
Brickwork	11 May 15:00	18 May 15:15
Roof	May 20 11:00	28 May 11:15
Windows	28 May 13:30	2 June 13:45

2 Here are three months of a calendar in 2023.

October						
Su	Mo	Tu	We	Th	Fr	Sa
1	2	3	4	5	6	7
8	9	10	11	12	13	14
15	16	17	18	19	20	21
22	23	24	25	26	27	28
29	30	31				

November						
Su	Mo	Tu	We	Th	Fr	Sa
			1	2	3	4
5	6	7	8	9	10	11
12	13	14	15	16	17	18
19	20	21	22	23	24	25
26	27	28	29	30		

December						
Su	Mo	Tu	We	Th	Fr	Sa
					1	2
3	4	5	6	7	8	9
10	11	12	13	14	15	16
17	18	19	20	21	22	23
24	25	26	27	28	29	30
31						

a A sports festival starts on 25 October. It is on for 3 weeks.
   On what date will it finish?

b A music festival starts on 29 November at 8 p.m.
   It will last for 5 days and 3 hours. When will it end?

c An art festival ends on the last Saturday in December.
   It is open for ten weeks. What date does it start?

Let's talk

The Rio Olympics began on 5 August, 2016.
How long ago was that? Be as exact as you can.

Quiz

1 Banko starts cooking at a quarter past 6 p.m.
   When he finishes, his clock shows:———→
   How much time did Banko spend cooking?

2 What will the date and time be 7 days from now?

Comparing and ordering fractions

Explore

Jin and Elok are playing a game of fraction snap.

Maths words

equivalent   denominator
compare   numerator
order

They each turn over
two cards at the same time.
The children win the cards if the two fractions are **equivalent**.
Will they win the cards this time?
Do any of the cards have the same **denominator**?
Use your specialising skills to play your own game of fraction snap.

Learn

It is easy to **compare** fractions with the same denominators.

$\frac{6}{8}$ has more equal parts than $\frac{5}{8}$, so we can write: $\frac{6}{8} > \frac{5}{8}$ or $\frac{5}{8} < \frac{6}{8}$

Let's compare the fractions $\frac{3}{4}$ and $\frac{1}{2}$.
What do you notice about the denominators this time?

We can use what we know about equivalent fractions to help us.
$\frac{1}{2} = \frac{2}{4}$, so we can now compare $\frac{2}{4}$ and $\frac{3}{4}$.

The denominators are now the same, but the **numerators** have changed. Which fraction is larger?

Equivalent fractions can also help us to **order** these fractions: $\frac{3}{4}$, $\frac{1}{2}$ and $\frac{5}{8}$
Can we write each fraction with the same denominator?

Practise

1 Compare each pair of fractions. Write **>** or **<** between them.

a $\frac{9}{10}$ and $\frac{7}{10}$

b $\frac{2}{3}$ and $\frac{6}{9}$

c $\frac{5}{6}$ and $\frac{2}{3}$

d $\frac{5}{6}$ and $\frac{11}{12}$

e $\frac{9}{10}$ and $\frac{15}{20}$

f $\frac{8}{10}$ and $\frac{4}{5}$

2 Order these fractions from smallest to largest.

$\frac{2}{5}$ $\frac{4}{5}$ $\frac{3}{5}$ $\frac{2}{5} < \frac{3}{5} < \frac{4}{5}$ a $\frac{3}{6}$ $\frac{1}{6}$ $\frac{2}{6}$ $\frac{4}{6}$

b $\frac{3}{4}$ $\frac{3}{8}$ $\frac{1}{2}$

c $\frac{9}{10}$ $\frac{4}{5}$ $\frac{7}{10}$

d $\frac{3}{6}$ $\frac{1}{3}$ $\frac{5}{6}$ $\frac{2}{3}$

e $\frac{5}{9}$ $\frac{2}{3}$ $\frac{2}{9}$ $\frac{1}{3}$

Remember to think about equivalent fractions!

Practise *(continued)*

 3 **Critique** these. Are they true or false? **Improve** any that are false.

a $\frac{7}{10} > \frac{6}{10}$

b $\frac{3}{4} < \frac{7}{8}$

c $\frac{1}{2} < \frac{7}{10}$

d $\frac{9}{10} > \frac{4}{10} > \frac{3}{5}$

e $\frac{2}{3} < \frac{3}{6} < \frac{6}{6}$

Try this

 2   5   8   10   4

Use four of these digits to make the ordering below correct.

$\frac{\square}{20} < \frac{3}{\square} < \frac{\square}{\square}$

Is there more than one solution?

I'm going to think about denominators that are easy to compare.

Let's talk

Elok compares two fractions. One fraction has the denominator 8. Elok uses equivalent fractions, so both fractions now have the same denominator.

Which two fractions could Elok be comparing? Find at least three possible pairs of fractions. Use the symbols < or > to show the relationship between the fractions.

Adding and subtracting fractions

Explore

Banko and Jin are showing different calculations with number lines.
What are the different calculations?
What are the answers?
What other calculations can the children show on these number lines?

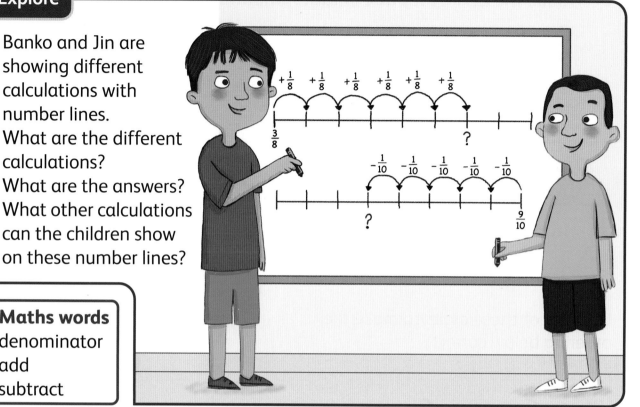

Maths words
denominator
add
subtract

Learn

It is easy to add or subtract fractions that share the same **denominator**.

We can show adding $\frac{4}{9}$ and $\frac{3}{9}$ in different ways.

| $\frac{1}{9}$ | $\frac{1}{9}$ | $\frac{1}{9}$ | $\frac{1}{9}$ | $\frac{1}{9}$ | $\frac{1}{9}$ | $\frac{1}{9}$ | | |

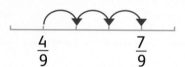

What happens when we **add** another $\frac{2}{9}$? And another $\frac{2}{9}$?

We can show **subtracting** $\frac{7}{9} - \frac{3}{9}$ in different ways.

| $\frac{1}{9}$ | $\frac{1}{9}$ | $\frac{1}{9}$ | $\frac{1}{9}$ | ✗ | ✗ | ✗ | | |

What do you notice?

Practise

1 Complete these. Draw number lines to help you.

a $\dfrac{4}{10} + \dfrac{3}{10}$ b $\dfrac{7}{10} + \dfrac{4}{10}$ c $\dfrac{5}{8} + \dfrac{2}{8}$

d $\dfrac{7}{10} - \dfrac{3}{10}$ e $\dfrac{10}{10} - \dfrac{5}{10}$ f $\dfrac{7}{8} - \dfrac{2}{8}$

2 Complete this question.

a Sort the calculations below into a table like this.
Use estimates to help you.

Answer is less than $\dfrac{1}{2}$	Answer is more than $\dfrac{1}{2}$ but less than 1	Answer is more than 1

$\dfrac{4}{6} + \dfrac{3}{6}$ $\dfrac{7}{8} - \dfrac{4}{8}$ $\dfrac{1}{4} + \dfrac{2}{4}$ $\dfrac{9}{10} - \dfrac{3}{10}$

$\dfrac{9}{12} - \dfrac{5}{12}$ $\dfrac{5}{7} + \dfrac{3}{7}$ $\dfrac{5}{9} + \dfrac{2}{9}$ $\dfrac{17}{20} - \dfrac{6}{20}$

b Make up another calculation to go in each section of the table.

3 Solve these problems.

a Guss uses $\dfrac{3}{10}$ of a bag of flour to make muffins.
He uses another $\dfrac{3}{10}$ of the bag to make cookies.
What fraction of the bag of flour is left?

b Pia runs $\dfrac{7}{8}$ km.
Sanchia runs $\dfrac{5}{8}$ km.
How much further did Pia run than Sanchia?

Try this

Find the missing fractions.

?	+	$\dfrac{3}{12}$	=	?
−		=		−
$\dfrac{3}{12}$	+	?	=	$\dfrac{5}{12}$
=		+		=
?	+	$\dfrac{1}{12}$	=	$\dfrac{6}{12}$

Introducing percentages

Explore

The children are talking about a recent times tables test.

Half of my times tables test was correct.

I got 50 % of my times tables test correct.

I got 100 % of my times tables test correct!

Guss got more of his test correct than Banko but less than Sanchia.
What fraction of his test could be correct?

Maths words
hundredth
percent
percentage

Learn

A whole is divided into 100 equal pieces called **hundredths**.

The word **percent** means **per hundred**.

One percent (1 %) is equal to $\frac{1}{100}$.

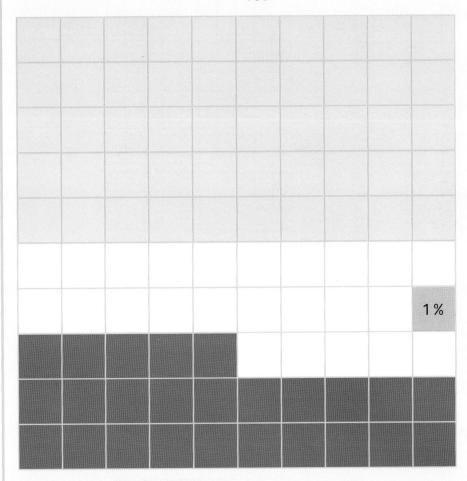

50 out of 100 small squares are shaded blue.

We write this as $\frac{50}{100}$.

This is the same as one half.

So 50 % is equal to $\frac{1}{2}$.

How many squares out of 100 are shaded pink?

Say this as a fraction and as a **percentage**.

Practise

1 What different percentages are shown here?

a
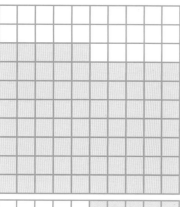

b
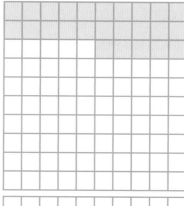

c
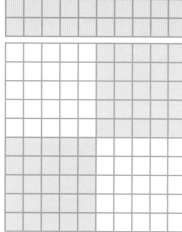

d
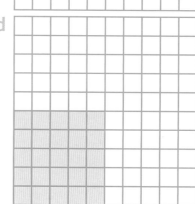

2 Write each percentage in question 1 as a fraction

with a denominator of 100: $\frac{\Box}{100}$

3 What fraction of each shape is shaded?
Write your answer as a fraction and as a percentage.

a
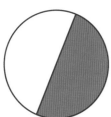

b
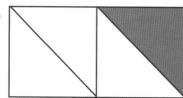

c
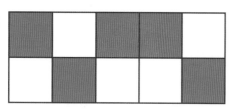

d
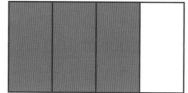

Try this

Elok cuts and then eats $\frac{1}{4}$ of an apple.

Pia eats $\frac{1}{2}$ of the same apple.

What **percentage** of the whole apple is left? **Convince** your partner.

Let's talk

Work together to complete these linking diagrams.
Explain how you know. What will you say or draw?

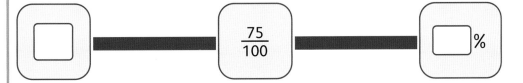

Quiz

1 Write the symbols **>** , **=** or **<** to make each statement true.

a $\dfrac{3}{4} \bigcirc \dfrac{15}{20}$ b $\dfrac{3}{4} \bigcirc \dfrac{7}{8}$ c $\dfrac{7}{10} \bigcirc \dfrac{4}{5}$

2 Order these fractions from smallest to largest.

a $\dfrac{6}{7}$ $\dfrac{3}{7}$ $\dfrac{4}{7}$ $\dfrac{5}{7}$ b $\dfrac{1}{5}$ $\dfrac{3}{10}$ $\dfrac{1}{10}$

c $\dfrac{7}{9}$ $\dfrac{1}{3}$ $\dfrac{2}{9}$ $\dfrac{2}{3}$

3 Complete these.

a $\dfrac{6}{10} + \dfrac{3}{10} = \bigcirc$ b $\dfrac{6}{10} + \dfrac{4}{10} = \bigcirc$

c $\dfrac{6}{10} + \dfrac{5}{10} = \bigcirc$ d $\dfrac{8}{9} - \dfrac{2}{9} = \bigcirc$

e $\dfrac{7}{9} - \dfrac{2}{9} = \bigcirc$ f $\dfrac{6}{9} - \dfrac{2}{9} = \bigcirc$

4 a Which shapes are 50 % shaded?

  b What fraction of the other shapes are shaded?
   Write these as percentages.

A   B    C

D   E

More coordinates

Explore

Look at this map.

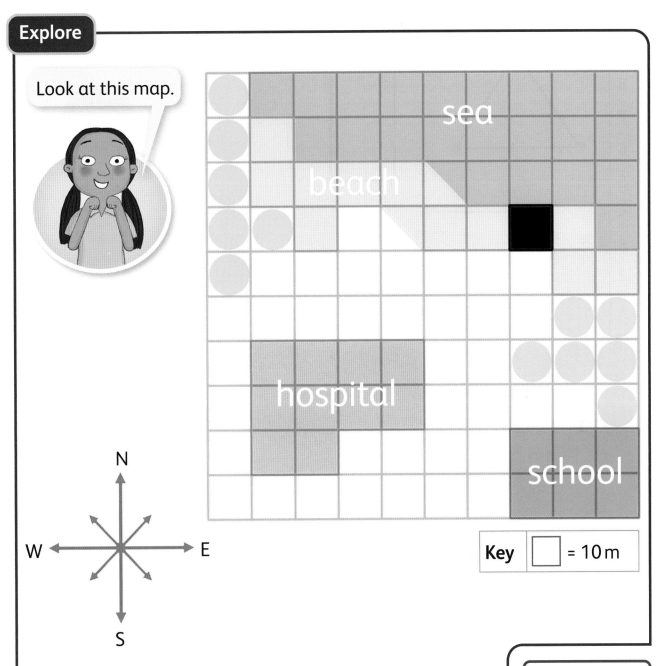

N

W   E

S

Key ☐ = 10 m

What is north-west of the school?
Estimate the distance from the school to the beach.
What is the perimeter of the hospital building?

Maths words
coordinates
x-axis
y-axis

Learn

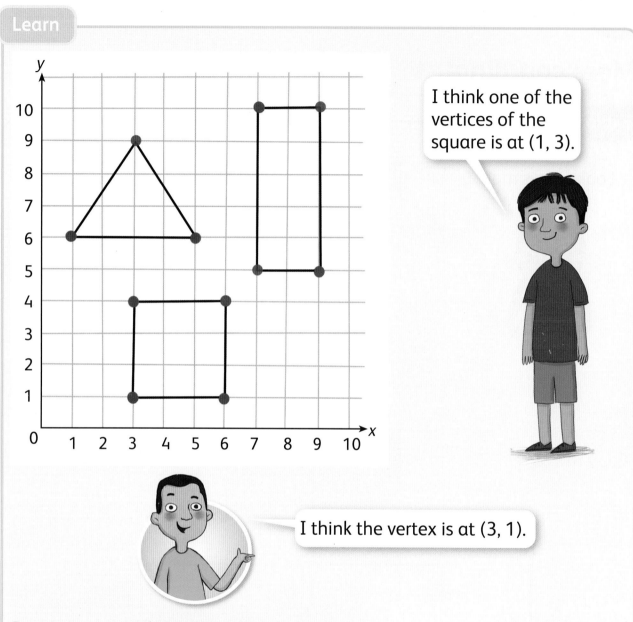

I think one of the vertices of the square is at (1, 3).

I think the vertex is at (3, 1).

Do you agree with Banko or Jin? Explain why.
What are the **coordinates** of the other three vertices of the square?
Remember to think about whether we read the **x-axis** or the **y-axis** first.

Practise

1 Look at the grid in Learn above.
   a Write the coordinates of the vertices of the triangle.
   b Write the coordinates of the vertices of the rectangle.

Practise *(continued)*

2 Two vertices of a square have been drawn. Write coordinates for the other two vertices. There are three possible solutions. Use your skill of **specialising** to try to find them all.

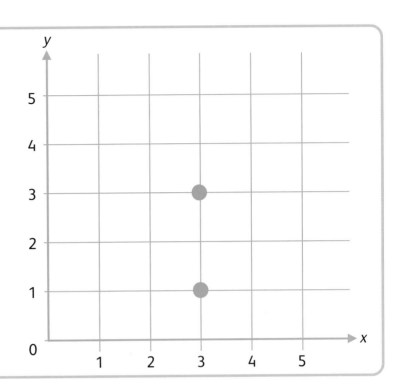

Try this

Copy this table.

Coordinates inside the square	Coordinates outside the square	Coordinates on the perimeter of the square

Use your skill of **classifying** to sort these coordinates into the three groups.

(5, 1)   (1, 5)
(5, 5)   (7, 6)
(6, 7)   (7, 7)
(2, 6)   (6, 2)

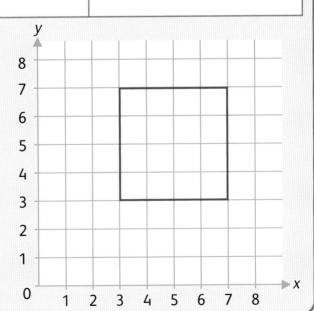

Reflections on a grid

Explore

Each shape is **reflected** along the dotted line, called the **mirror line**.
We have completed the first shape. It is a **symmetrical** shape with one line
of symmetry. It has three vertices. Predict how many vertices are in each
completed shape. Then check by drawing or using a mirror.

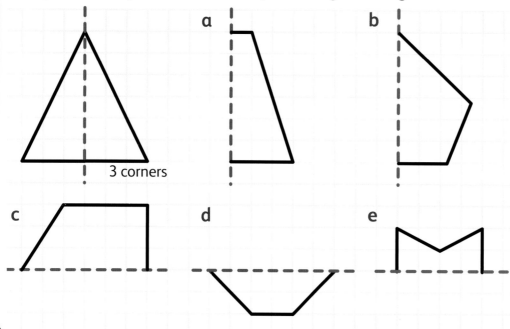

3 corners

Learn

Each shape has been reflected incorrectly.
Critique them and explain the mistakes.

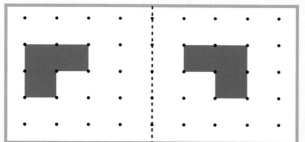

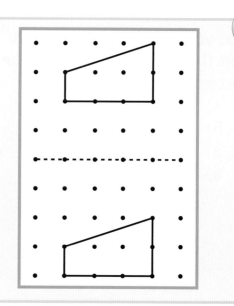

Use dotty paper or a square grid to show
these reflections correctly and improve them.

Practise

1 Use dotty paper or a square grid to reflect each shape.

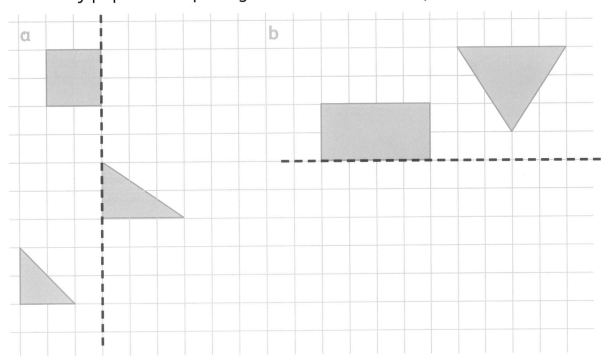

2 What is the smallest number of squares you need to add to make each shape symmetrical across its mirror line?

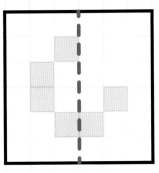
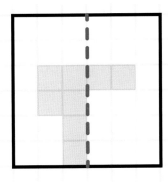
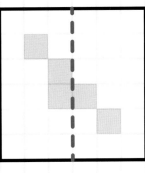

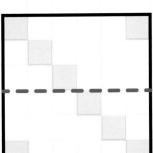
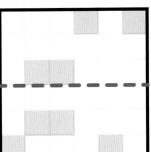
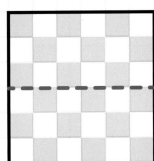

Try this

Reflect each shape in the mirror line. Write the coordinates of the new vertices of each shape.

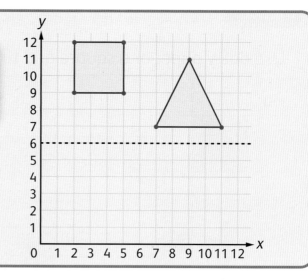

Quiz

1 These are two vertices of a right-angled triangle. Give the coordinates to complete the triangle.

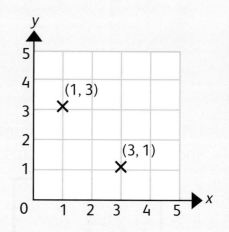

2 Reflect each shape correctly on grid paper.

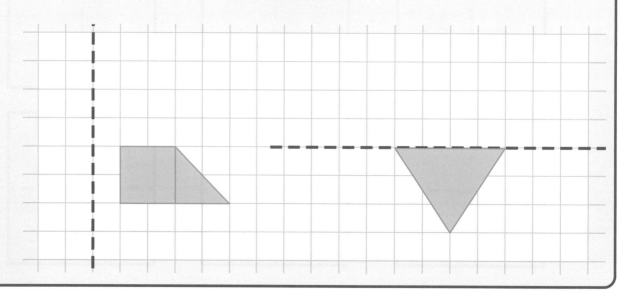

Review

Units 13–18

1 Draw arrays to show which of these numbers are square numbers.
  a 12   b 9   c 16   d 20

2 Write the first five numbers that are greater than 400 each time.

  a [Multiples of 2]   b [Multiples of 5]   c [Multiples of 10]

3 a Explain the difference between these four charts:
    a pictogram, a tally chart, a dot plot and a bar chart.
  b When would you choose to use each chart?

4 Find the missing numbers.
  a $43 \times 10 =$ ☐   b $43 \times$ ☐ $= 4\,300$

  c $5\,400 \div 100 =$ ☐   d $540 \div$ ☐ $= 54$

5 Copy and complete.
  a $96 \div 6 =$ ☐   b $91 \div 7 =$ ☐

  c $26 \div 3 =$ ☐   d $78 \div 5 =$ ☐

6 A festival began on 8 February at midday.
  It ended 3 days and 3 hours later. When exactly did it end?

7 Write >, = or < to make each statement correct.

  a $\dfrac{3}{4}$ ☐ $\dfrac{6}{8}$ b $\dfrac{9}{10}$ ☐ $\dfrac{5}{10}$ c $\dfrac{2}{3}$ ☐ $\dfrac{10}{15}$ d $\dfrac{2}{5}$ ☐ $\dfrac{5}{10}$

8 Which is the odd one out each time?
  a $\dfrac{1}{2}$, 50%, $\dfrac{5}{100}$, $\dfrac{50}{100}$ b 25%, $\dfrac{2}{5}$, $\dfrac{1}{4}$, $\dfrac{25}{100}$ c $\dfrac{3}{4}$, 75%, $\dfrac{7}{5}$, $\dfrac{75}{100}$

9 Draw a coordinate grid and mark these points:
  A at (2, 3) and B at (3, 2).

2D shape a flat shape

3D shape a solid shape

12-hour (clock) uses the numbers 1 to 12 for the hours, and uses a.m. and p.m. to show if the time is in the morning or at night

24-hour (clock) shows time across a day as 24 hours, using the numbers 00:00 to 23:59 (midnight is at 00:00)

a.m. after midnight and before midday

p.m. after 12 noon, in the afternoon

A

acute an angle that is less than 90 degrees

acute angles

add to find a sum or total

addition a calculation of the sum of two numbers or things

angle the amount of turn between where two lines meet

area the amount of space that a flat surface or a shape covers

array a rectangular arrangement of quantities

array

associative when we add or multiply three or more numbers, the sum (or the product) is the same, regardless of the grouping of the addends (or the factors)

B

bar chart a chart that uses bars to show the relationship between groups of information

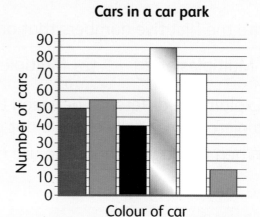

Cars in a car park

Number of cars

Colour of car

C

calendar pages or tables that show the days, weeks and months of a year

Carroll diagram used to sort items according to two groups

centimetre a unit for measuring the length of something; 100 cm = 1 m, 10 mm = 1 cm

1 cm

certain when it is absolutely likely that something will happen; when we are sure about something

chance when something might or may probably happen

chart information in the form of a table, graph or diagram

collect to group or put together with other objects

commutative when an operation (+ and ×) can be done in any order and the result stays the same, for example: 6 + 4 = 10 and 4 + 6 = 10; 3 × 4 = 12 and 4 × 3 = 12

compare to note similarities and differences

compound shape a shape that can be divided into one or more basic shapes

coordinate a pair of numbers or values to show an exact position, for example, on a graph

D

data information

decide make a decision or make up one's mind about something

decompose (partition) to break up or split numbers into their place value parts, for example:

degree a unit of measurement of angles

denominator a number in a fraction below the line, also called the divisor $\longrightarrow \frac{3}{4}$

diagonal a straight line that joins two opposite corners of a polygon

difference subtract one thing from another, for example, the difference between 392 and 250 is 142

digit one of the written signs we use to represent the numbers 1 to 9

direction course or path along which something or someone moves

divide to find how many times a number goes into another number

divisible when a whole number can be divided by another whole number without a remainder

division separating something into equal parts

dot plot a chart used to collect and present information

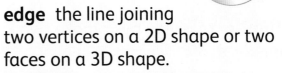

0 1 2 3 4 5

double twice as many or as much

duration the length of time of an event from start to finish or beginning to end

E

east the direction 90 degrees clockwise from north

edge the line joining two vertices on a 2D shape or two faces on a 3D shape.

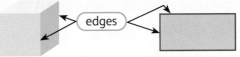

edges

equal the same value as

equivalent the same value as

estimate to try to guess the value, size, amount, cost, speed, …, of something without calculating it exactly

even (number) a number that can be divided exactly by two

experiment a test conducted to study something

explain make something clear by giving reasons or details

F

face a surface of a solid shape

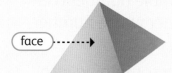

factor a whole number that divides into another whole number exactly

factor pair a set of two numbers we can multiply to get a product

fraction a part of a whole, for example: $\frac{1}{2}$ of 6 = 3

frequency how often something happens

G

group to gather/collect or put together

H

half/halve to divide by 2; make two equal parts

horizontal level to the ground, at a right angle to the vertical

hundreds ten groups of ten

hundred thousand(s) a hundred groups of 1 000; ten times larger than 10 000; one more than 99 999; the place value that shows a number multiplied by 100 000

hundredth when a whole is divided into 100 equal parts, each part is one hundredth $\frac{1}{100}$ of the whole.

I

impossible when something has no chance of happening

information for example, facts or data from a survey or learning about something

interpret to explain the meaning of

inverse the opposite of an operation, for example, subtraction is the inverse of addition

investigate to find out more about or research something to find out the facts or the truth

L

likely something that might well or will probably happen

line of symmetry when a straight line can be imagined to be drawn through an object or shape to divide it into two parts so that one side looks like a mirror image of the other side

M

maybe might happen or has a chance of happening

millimetre a unit for measuring length; 1 000 millimetres = 1 metre

million(s) a hundred groups of 10 000; ten times as large as 100 000; one more than 999 999; the place value that shows a number multiplied by 1 000 000

mirror line the line that marks how both sides of something the same, so that one side looks as though it has been flipped

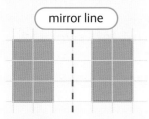

mirror line

month there are 12 months in a year: January, February, March, April, May, June, July, August, September, October, November, December

multiple a whole number that can be divided equally by another whole number without a remainder

multiplication a way of calculating the product of two numbers; it can also sometimes be described as adding a number to itself a specific number of times

multiply (whole numbers) calculate the product of two numbers

N

near multiple (of 10) the multiple of 10, 100, 1 000, …, that a number is closest to

nearest multiple (of 10) the multiple of 10, 100, 1 000, …, that a number is closest to, for example, 1 300 is the nearest multiple of 100 to 1 295

negative number a number that is less than zero (0)

net a pattern that you can cut and fold (or imagine cutting and folding) to make a model of a 3D shape

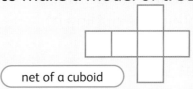

net of a cuboid

north the direction in which a compass needle points

number line a line with points that represent numbers

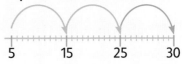

numerator the number above the line in a fraction; in the fraction 2 __ 5 , the 2 is the numerator

O

obtuse an angle that is greater than 90 degrees, but less than 180 degrees

obtuse angles

odd numbers that are not exactly divisible by 2, for example: 1, 3, 5, 7

order an arrangement of objects or numbers

P

parallel lines that are always the same distance apart and never meet

part a piece or a fraction of a whole

percent or percentage part of a whole; 'percent' is per hundred; 25 % is equal to $\frac{25}{100}$

perimeter the length or distance all the way around the outline of a 2D shape

place value the value every digit has in a number, for example, 2 453 has 3 ones, 5 tens, 4 hundreds and 2 thousands

1 000s	100s	10s	1s
2	4	5	3

place value chart

polygon a flat (2D) shape with three or more straight sides

positive number a number that is more than zero

prediction a guess that you provide with reasons

probability the likelihood of an event happening: likely, unlikely, certain or impossible

product the result after multiplying numbers together

R

random not in a pattern or particular order

recursion (rule) the rule describing a sequence and the relationship between the terms. For example, in the sequence 2, 4, 6, 8, … the recursion rule is 'add 2' because the next term in the sequence can be found by adding 2 to the previous term

reflect to create a mirror image

reflection the mirror image of something; the image is 'flipped' but is the same size and shape as the original

regroup exchange for something with the same value, for example, regroup 10 ones to make 1 ten; regroup 1 365 as 1 300 + 65 or 1 200 + 160 + 5

remainder the amount left over after division

right angle an angle that is 90 degrees

right angle

round increase or decrease a number to the nearest multiple of ten, hundred, thousand, … (for example, 3 452 rounded to the nearest hundred is 3 500)

rule an accepted or usual way or method of doing or following something

S

scale a system according to which things are measured

scale an object that is used to measure weight

balancing scale, used to measure weight

sequence a list of numbers in which

each number is obtained according to a specific rule

sort put in order or arrange in groups

south the direction 180 degrees from north or 90 degrees clockwise from east

square number is a number that is formed by multiplying the same number with itself, for example, 25 is a square number because it is formed by 5 × 5

straight a line or road that follows the same direction; has no curves or bends

subtract to take away one number from another number, or find the difference between two numbers

subtraction taking one number or amount away from another number or amount, or finding the difference between two amounts

sum to add to find a total

symbol a shape or sign used to represent something

symmetrical when each half is exactly the same

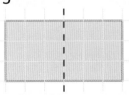

symmetry when a straight line can be drawn through an object or shape to divide it into two parts so that the one side looks like a mirror image of the other side (see also *line of symmetry* and *mirror line*)

T

tens the place value that shows a number multiplied by 10

term number sequences follow a rule that connects each value within them; these values are called terms

tessellate when repeating geometric shapes, such as tiles, fit together so that there are no gaps (spaces) and could continue doing so forever

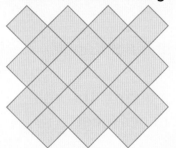

thousands the place value that shows a digit multiplied by 1 000

timetable a chart showing the departure and arrival times; a schedule or program of events with starting and ending times

total the answer to an addition calculation

triple three times as many

turn to rotate (turn from its centre) or change position

turn 90° clockwise

U

unlikely something that has little chance of happening

V

Venn diagram a diagram with circles to show sets

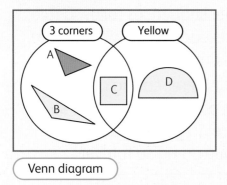

Venn diagram

vertical upright, at a right angle to the horizontal

vertices (plural for vertex) the point where two sides or lines meet

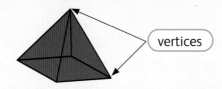

vertices

W

west the direction a three-quarter turn or 270 degrees from north

whole something that is complete; it has all its parts

X

x-axis the horizontal axis on a graph

Y

y-axis the vertical axis on a graph

Z

zero nothing; zero is neither positive nor negative

Thinking and Working Mathematically (TWM) skills vocabulary

characterising identifying and describing the mathematical properties of an object

classifying organising objects into groups according to their mathematical properties

conjecturing forming mathematical questions or ideas

convincing presenting evidence to justify or challenge a mathematical idea or solution

critiquing comparing and evaluating mathematical ideas, representations or solutions to identify advantages and disadvantages

generalising recognising an underlying pattern by identifying many examples that satisfy the same mathematical criteria

improving refining mathematical ideas or representations to develop a more effective approach or solution

specialising choosing an example and checking to see if it satisfies or does not satisfy specific mathematical criteria